This book is to be returned on or befor
the last date stamped helc

1991

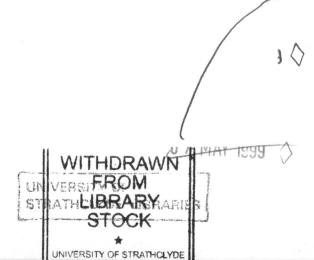

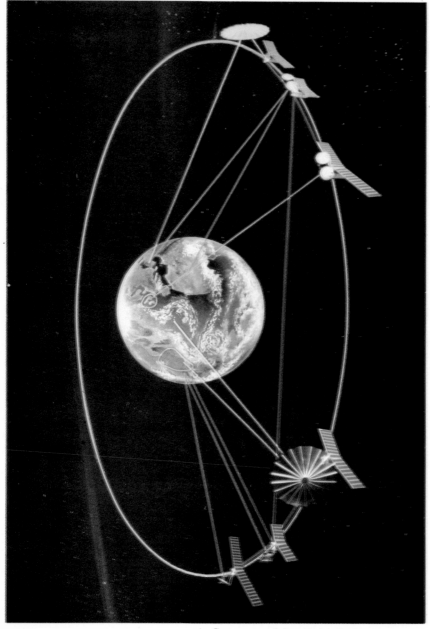

Sponsored by British Aerospace
**Big Communicators Servicing National and International
Regions for Transfer of Traffic**

Introducing
Satellite Communications

G B Bleazard

PUBLISHED BY NCC PUBLICATIONS

British Library Cataloguing in Publication Data

Bleazard, G.B.
  Introducing satellite communications.
  1. Artificial satellites in telecommunication
  I. Title
  621.38'0422   TK5104

ISBN 0-85012-471-9

Library of Congress Cataloguing in Publication Data

Bleazard, G. B.
  Introducing satellite communications.
  "A Halsted Press book".
  Bibliography.
  Includes index.
  1. Artificial satellites in telecommunication.
  I. Title
  TK5104.B54 1985   384.54'56   85-12228

ISBN 0-470-20228-9

© THE NATIONAL COMPUTING CENTRE LIMITED, 1985

First published in 1985 by:

NCC Publications, The National Computing Centre Limited, Oxford Road, Manchester, M1 7ED, England.

Typeset in 10pt Times Roman and printed by UPS Blackburn Limited, 76-80 Northgate, Blackburn, Lancashire, and bound by John Sherratt and Son Limited, 2 Gloucester Street, Manchester.

ISBN 0-85012-471-9

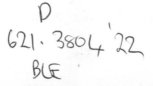

Acknowledgements

Many organisations and individuals have contributed, either directly or indirectly, to the studies which resulted in this book.

First of all we wish to acknowledge the assistance received from the following organisations and their staffs (those who were visited, who arranged the visits and made time available for discussion; and who provided descriptive material relating to their products and activities):

In the USA

IBM Corporate Communications Headquarters	New York
AT&T Long Lines Division	New York
Merrill Lynch & Co Inc	New York
Citibank	New York
The Wall Street Journal	New York
SBS (Satellite Business Systems)	McLean, Virginia
COMSAT World Systems	Washington DC
INTELSAT	Washington DC
Video Star Connections Inc	Atlanta
Harris Corp, Satellite Communications Division	Melbourne, Florida
The Kennedy Space Centre	Florida
National Aeronautics and Space Administration (NASA)	Washington
Atlantic Richfield Corporation	Los Angeles
TRW Space & Technology Group	Los Angeles
Oak Industries	San Diego
Bank of America	San Francisco

In Europe

BTI (British Telecom International)
The French Ministry of Posts, General Telecommunications Department
British Aerospace, Space & Communications Division
Marconi Space & Communications Systems, Space & Microwave
    Division
Plessey-Scientific Atlanta
Aerospatiale, France
EUTELSAT
ESA (European Space Agency)
ITU (International Telecommunication Union), Geneva
CCITT, Geneva

Many individuals have helped in one way or another but particular
thanks are due to the following:

John Hardy   –   British Telecom International
Dave McGovern   –   British Telecom International
Kevin Hodson   –   Communications Systems Research
Larry Blonstein   –   British Aerospace Ltd
David Gregory   –   British Aerospace Ltd
Richard Barnett –   British Aerospace Ltd
Ivor Knight   –   COMSAT World Systems
Jack Dicks   –   INTELSAT
Sebastian Lasher –   INTELSAT

Special thanks are also due to an NCC colleague, Brian West, for
accompanying the author on a fact-finding tour in the USA, and for
commenting on the draft.

The book is illustrated with diagrams, photographs and other material
drawn from a variety of sources. These are acknowledged in the appro-
riate places in the text. We are particularly grateful to the following
companies for sponsoring the colour photographs (acknowledgements
are given on the individual plates):

    British Telecom International
    British Aerospace
    Marconi

We also thank David Shorrock of SCICON (Scientific Control Systems) Ltd and On-Line Conferences Ltd, for permission to quote freely

from a published paper on the economics of business satellite services in the United Kingdom. This paper only became available when the drafting process was in its final stages. This was a fortunate coincidence indeed, because it enabled a greater measure of quantification to be introduced into the discussions and results of Chapter 11 than would otherwise have been possible.

We are grateful to Prentice-Hall for permission to reproduce Figures 4.2, 5.5, 5.7, 6.5 and 9.4 which are adapted from James Martin's *Communications Satellite Systems*.

The reproduction in this book of material taken from publications of the International Telecommunication Union (ITU), Place des Nations, CH-1211 GENEVA 20, Switzerland, has been authorised by the ITU.

Thanks are also due to the NCC typists who coped, often in difficult circumstances, with a considerable body of work.

Finally, the Centre acknowledges with thanks the support for this project by the Avionics and Electronics Requirements Board of the Department of Industry.

Contents

Page

Acknowledgements

1 Introduction   17

2 Origins and Basic Principles   23

   Origins   23
   Basic Principles   24
      General Features   24
      The Space Segment   28
      The Earth Segment   30
   Properties of Satellite Communications Systems   34
      The Broadcast Property   34
      Geographical Flexibility   35
      Distance-Insensitive Costs   37
      Transmission Capacity and Speeds   37
      Transmission Delay and Echo   38

3 Evolution of Satellite Communications   41

   Introduction   41
   The Experimental Phase   41
   The First Geostationary Satellite   43
   Expansion and Growth   43
   International Services   46
   INTELSAT Series   47
      INTELSAT I (Early Bird)   47
      INTELSAT II   48

INTELSAT III   48
INTELSAT IV   49
INTELSAT IVA   49
INTELSAT V   49
INTELSAT VI   50
Emergence of National and Regional Services   50
The Impact of Satellite Broadcast Television   53
Mobile Satellite Communications   54
The Emergence of Small-Dish Systems   55
Direct Broadcast Television   56
Specialised Business Services   57
Technological Trends in Satellite Design   59

4 The Satellite Orbit   67

Introduction   67
Orbital Dynamics   67
Orbital Characteristics   68
    Orbital Shape   68
    Types of Orbit   69
    Advantages/Disadvantages of Geostationary Orbit   72
Satellite Spacing and Orbital Capacity   73
Angle of Elevation   75
Eclipses   80
Launching and Positioning   82
    Launch Vehicles   83
    Launch Procedures   89
Satellite Drift and Station Keeping   92

5 Satellite Construction   95

Introduction   95
The Antennae Subsystem   98
    General Principles   98
    Antennae Performance: Gain and EIRP   100
    Types of Antennae   101
    Steerable Antennae   106
    Area of Coverage: Footprints and Beam Shaping   109
    Frequency Re-use   113
The Transponder Subsystem   113
    A Typical Transponder Subsystem   115

High Power Amplifiers 117
Power Supply System 118
Solar Energy 118
Two Types of Solar Array 119
Stabilisation and Attitude Control 121
The Two Types of Stabilisation System 121
Position Measurement 122
Positioning Accuracy 123
Other Subsystems 124
Tracking, Telemetry and Command (TTC) 124
Thrust Subsystem 125
Reliability 125
Component Selection and Testing 125
Construction and Assembly 125
Component Redundancy 126
Satellite Back-up 126
Satellite Life-span 126
The Space Environment 127
The Beneficial Effects 127
Effect on Materials 128
Lubrication 128
The Heat Transfer Problem 129
Solar and Other Radiation 129
Space Debris: Natural and Man-made 130

6   The Transmission Link: Structure and Performance 133

Introduction 133
Principles 135
Antennae Beam Frequency Assignments 135
Channel Subdivision 135
Transmit/Receive Chains and Path Interconnection 135
Earth Station Frequency Assignments 138
The Frequency Plan 138
Example: The ECS-1 Satellite 138
Frequency Considerations 141
Operating Frequency Selection 141
Frequency Allocation and Co-ordination 142
Choice of Frequency 145
Interference 14⁻

Transmission Losses 149
  Free Space Loss 149
  Atmospheric Losses 150
Noise 152
  General 152
  The Measurement of Noise 153
  Noise Sources 154
Combatting Signal Loss and Noise 156
Transmission Link Performance 157
  The Link Budget 157
  Information-Transmission Capacity 160
  Satellite Channel Bit Error Rate (BER) 162
Overall System Performance 163

7 The Earth Segment: Transmission, Terrestrial Infrastructure 167

Introduction 167
Transmission Principles 167
  Modulation and Modulation Techniques 167
  Analogue and Digital Transmission 174
  Multiplexing 176
Terrestrial Transmission Path and Baseband Signal
  Processing 184
  Analogue-to-Digital Conversion 185
  Digital Speech Compression 187
  Video Compression 187
  Digital Speech Interpolation (DSI) 188
  Forward Error Correction (FEC) 189
  Encryption 190
Terrestrial Access Arrangements 192
  Shared and Private Earth Terminals 193
  The User Interface 194
  Terrestrial Network Access 195
  Network Interfacing and Transmission Standards 196
  Service Access Requirements 197

8 The Space Segment: Access and Utilisation 201

Introduction 201
The Earth Terminal 201

General Features   201
Types of Earth Station and Earth Station
   Standardisation   203
Earth Terminal Performance   204
The Earth Terminal Transmission System   204
Space Segment Access Methods   208
  Classification of Multiple-Access Methods   209
  Frequency Division Access (FDMA)   210
  Single Channel Per Carrier (SCPC) Systems   214
  Time Division Multiple Access (TDMA)   216
Comparison of FDMA and TDMA Techniques   223
  FDMA Techniques   224
  TDMA Techniques   225
  Service Implications   226
Assignment Methods   226
  Pre-Assignment   228
  Demand Assignment Multiple Access (DAMA)   228
  The SPADE Demand Assignment System   229
  TDMA Demand Assignment   230
  Random Access   231
  Miscellaneous Access Techniques   232
System Control   233
  Centralised Control   234
  Decentralised Control   234

9 Applications Impacts: Echo and Delay Considerations   237

Background   237
Speech   238
  Delay Effects   238
  Speech Echo   240
Data Transmission   243
  Protocol Functions   243
  Performance and the Satellite Transmission
    Environment   246
  Satellite Delay Implications   248
  Protocol Alternatives   251
  Implementation Considerations   262
  Conclusions and Recommendations   264

10   The Role and Application of Satellite Communications   267

Satellite Communications in Context   267
   Competition from Terrestrial Services   268
   Geographical Factors   268
   Social and Economic Factors   269
   The Role of Television   270
   Political Factors   270
   The Regulatory Environment   271
   The Complementary Role of Satellite Communications   274
   Supply Flexibility   276
   Early Availability of Wideband Digital
      Transmission Services   278
   Capacity Flexibility   278
   High Interconnection Potential   278
   Distance-Insensitive Costs   279
Applications and Application Opportunities   279
   Traditional and Miscellaneous   280
   Commercial and Allied Applications   286

11   Satellite Business Services in Europe and UK   297

Introduction   297
UK Service Plans   298
   SatStream North America   298
   SatStream Europe   298
   SatStream Offshore   298
Service Presentation   302
Space Segment Access and Capacity Assignment   303
SatStream Europe Facilities   303
   Earth Station Access   303
   Transmission Rates   304
   Space Segment Channel Configurations and Utilisation
      Arrangements   304
Economics of Satellite Business Services   304
   A US Example   305
   Satellite versus Terrestrial Services in Mainland UK   308
   SatStream Services within Continental Europe   313
Conclusions   314

worth noting

Epilogue: Satellite Communications in Perspective   317

Bibliography   321

Appendix

   A   A Note on Decibels   323
   B   Glossary of Terms   325
   C   List of Acronyms   335

Index   339

1 Introduction

The age of satellite communications began during the night of 10-11th July 1962, when television pictures were first flashed across the Atlantic via the Telstar 1 satellite. From that time the development and exploitation of satellite communications has progressed at an unrelenting pace.

Only three years after Telstar, the first INTELSAT satellite was launched. Today INTELSAT has approximately 17 satellites strung above the equator like a string of beads, conveying about two thirds of the international telecommunications traffic across the oceans of the world, and relaying news and pictures of events to the screens of a vast population of television viewers. For example, by the time this book reaches the reader, the Los Angeles Olympic Games will already be a fading memory, but it is estimated that the INTELSAT system would have enabled the Games to have been received live by around 600 million TV receivers throughout the world, whether in the advanced countries or elsewhere.

The relaying of speech and television programmes was the earliest application of satellite communications, and continues to be its major role today. Indeed it has become so commonplace that we tend to take it for granted as we watch our television screens or make long-distance telephone calls. About seven years ago a number of factors converged to influence the nature of satellite communications. These factors included: significant improvements in cost and performance of satellite systems; the development of the small-dish earth station; the growing feasibility of employing digital transmission technology; a recognition of its many advantages; and the convergence of communications and computing technologies which underlies the Information Technology Revolution. Together these have combined to focus attention on a whole new range of

17

application opportunities for satellite communications, many of which were technically beyond the capability of the current terrestrial transmission services. Examples include: business applications (such as video conferencing and high-speed facsimile); information distribution and retrieval; and the direct broadcast of TV services to individual home antennae.

Almost from the very beginning, satellite services have been more widely and rapidly exploited in the USA and Canada than anywhere else. In addition to using the INTELSAT satellites for international transmission, these countries were quick to establish their own domestic satellites, and to exploit them for a variety of applications. In particular, they were used for broadcasting television to the already well-established cable networks, and were also made directly available to commercial organisations for telephony and data transmission.

In sharp contrast, European countries have been relatively slow to proceed along this path. However, Europe as a whole handles a significant quantity of INTELSAT traffic and has, over the years, carried out ambitious research programmes on the use of satellites for scientific communications purposes. (There are a number of reasons for the disparity between Europe and North America; these are discussed in Chapters 3 and 10.)

Following the decision, several years ago, by the European Telecommunications Administrations to establish a European satellite communications service, and parallel initiatives by individual countries to establish national satellite services, Europe has now embarked upon a programme of rapid expansion. (The essential features of this programme are briefly outlined in Chapter 11.) In 1984 direct customer access to satellite services – whether through private or shared earth stations – first became available. The kinds of application which will be supported include those referred to above, as well as traditional telephony and TV distribution.

This constitutes the primary reason for publishing this book. Until recently, satellite communications was generally somewhat remote from the end user; it was largely taken for granted, despite the fact that it already affected the lives of most people. Now it is set to have an even more widespread impact. It is destined to play an increasingly significant role in the national and international communications infrastructures in the years that lie ahead.

The approach of the present book falls between the popular and the technical – hopefully, closer to the former than the latter. We have tried to explain the concepts rather than the underlying technical details; where necessary, brief explanations of technical apparatus are provided.

As with any new brand of technology, satellite communications has sponsored an abundance of jargon and acronyms. These are defined at the appropriate places in the text, and glossaries of technical terms and acronyms are also provided to assist the reader. Even the person who experiences some difficulty with the technicalities should be able to read the book with profit.

A bibliography has been included. Rather than confuse the reader with a plethora of material we have provided a short list of references. Several of these have been found useful during the preparation of this book, and the reader may also find them equally helpful, either as background reading or to point in further directions. Preceding the bibliography proper, there is a short list of references covering the basic principles of telecommunications and data transmission – for readers who may lack familiarity with these subjects, or who may want to refresh their memories.

A brief glance through the contents pages will indicate the range of topics covered in this book. Chapter 2 provides a bird's eye view of the subject, covering both the historical origins and basic principles, with Chapter 3 presenting a more detailed historical perspective. Chapters 4 and 5 are concerned primarily with the space engineering aspects, covering such topics as the choice of satellite orbit, satellite construction, and satellite launching procedures. The mechanics of a satellite launch and the awesome grandeur of space continue to capture the public imagination. We have assumed that the reader would be no less interested in learning a little more about these aspects. The construction and placing in orbit of a satisfactorily functioning communication satellite is a triumph of technology and human ingenuity. In this field the engineering problems are formidable and require novel and exotic solutions.

Chapters 6 onwards move progressively from a consideration of the satellite transmission link as a conveyor of information, to the impact of the transmission link on applications and performance and then to a discussion of the role and application of satellite communications. Thus, Chapter 6 embraces topics such as frequency selection and coordination, and link performance. Chapter 7 introduces basic transmission principles, relevant to terrestrial access and other topics, and also to matters discussed in Chapters 8 and 9. Chapter 8 deals with the techniques used for sharing a satellite's capacity between many users, together with the relevant access methods. These topics are a frequent cause of confusion, particularly in popular accounts.

Two of the unique properties of a satellite link – namely, echo and delay – have given satellite communications an aura of notoriety. In Chapter 9 we give an account of their implications, the circumstances in which their effects may be severe, and how such effects may be ameliorated or eliminated. Chapter 10 discusses the role of satellite communications in general and within the specific European and UK context, and then discusses particular satellite services. Chapter 11 focuses on the United Kingdom. We review the current service plans and then, using examples, evaluate the economic benefits for business users. These evaluations are based upon a combination of published tariffs and estimated costs where tariff information is unavailable.

The pace of development of satellite communications continues unabated, and we indicate some of the major on-going trends which are expected to reach fruition in the short- to medium-term. However, terrestrial communications is also experiencing a parallel rapid development. As a consequence, in contrast to the earlier position, satellite communications is now facing increasing competition from these new and powerful terrestrial services. To conclude the book, Chapter 12 briefly reviews the evolving role of satellite communications within this broader perspective. Apart from this limited excursion into future possibilities, the book concentrates on basic principles, and the proven technology utilised by the majority of current operational services.

There will be many people who wish to find out more about the technology and application of satellite communications – either out of general interest, or because the nature of their job demands such an appreciation. For instance, at a time when communications technology generally is experiencing such an unprecedented growth, commercial enterprises will need to consider the implications of satellite communications when planning their corporate communications strategies. This book is designed to fulfil these needs.

A secondary reason, related to the first, is the shortage of suitable texts which provide a comprehensive overview of the subject and which also relate to the European scene. (The valuable book by James Martin, for instance – see Bibliography – is addressed primarily to an American audience.)

The original intention was to address the book to communications specialists and to those responsible for corporate communications, but it was soon realised that it could be of interest to a wider audience. This extended readership would certainly include workers in the various branches of Information Technology, who need information about satellite communications. The book may also supply communications specialists with a breadth of perspective not easily acquired in the working environment. Finally, with the proliferation of personal computers and the widening public awareness of Information Technology, it is thought that the book might also interest other lay persons, for example the growing band of computer hobbyists.

2 Origins and Basic Principles

ORIGINS

In 1945, the English science-fiction writer Arthur C. Clarke, writing in *Wireless World*, proposed a global communications system utilising manned space stations orbiting the earth at the same rate as the earth's rotation:

> "An 'artificial satellite' at the correct distance from the Earth would make one revolution every 24 hours; ie it would remain stationary above the same spot and would be within optical range of nearly half the Earth's surface. Three repeater stations, 120 degrees apart in the correct orbit, could give television and microwave coverage to the entire planet".

In 1955, J. R. Pierce, the Scientific Director of the US Bell Laboratories, expanded upon Clarke's theories. In particular, he defined the parameters for the utilisation of space satellites for communications.

Whilst the history of space satellites is relatively recent, the history of rocket propulsion goes back much further. The Chinese, having invented gunpowder many centuries ago, employed it in sky rockets on festive occasions. The discovery in Europe that steam jets could provide motive force, coupled with an appreciation of Newton's Laws of Motion, provided an understanding of the principles that underlie rocket propulsion.

Major theoretical contributions were made by the Russian mathematician Konstantin Tsiolkovsky who carried out extensive researches into the dynamics of rocket propulsion. He explored, for example, the conditions to be satisfied for a rocket to escape from the earth's gravitational field; and he also proposed, in 1903, that rockets could be used for the exploration of space.

The first significant experimental work on rocketry was carried out in the 1920s by the American Robert H. Goddard. His rockets did not travel very far, and most of the principles he followed, although interesting, have been superseded. However, he was the first to experiment with liquid fuel, using the mixture of gasoline and oxygen that was to be exploited by the German V2 rockets of World War Two. (A liquid-fuel mixture continues to serve as the primary propellant on present-day rockets.) The German V1 rocket was little more than an aerodynamic flying bomb of limited accuracy. The V2, a far more massive device, was designed to operate with greater accuracy and outside the earth's atmosphere. At the end of the war, the German scientists were dispersed, some moving to Russia and others to the US, to make a significant contribution to the rocket development and space exploration programmes of their adopted countries. In particular, Wernher von Braun, deputy director of the wartime German military programme, was to play an important role in the US development of the very powerful first-stage rockets which were needed to launch satellites to increasingly higher altitudes.

BASIC PRINCIPLES

General Features

The quotation from Clarke (p 23) highlights the two key attributes of most current and planned communications satellites: the *geostationary orbit,* and the satellite acting as a *relay* or *repeater station*.

If a satellite at a particular height above the equator moves at the correct speed, it will travel once around the earth in the same time that the earth takes to complete one rotation. If it also travels in the same direction as the earth's direction of rotation it remains over the same point on the equator and appears stationary to an observer on the ground. Hence the expression 'geostationary orbit'. (These relationships are illustrated in Figure 2.1.) Positioned 22,500 miles (36,000 kilometres) from the earth, the satellite travels at a speed of 6,879 miles/hour and completes its orbital journey of 264,000 km in approximately 24 hours.

A geostationary satellite is visible from slightly less than one-half of the earth's surface. In fact, three satellites in this orbit, located at the vertices of an equilateral triangle with sides of 88,000 km, are sufficient to give total coverage of the globe, excluding the polar regions. It is this vast coverage of the earth's surface which permits, for example, direct communication between England and Japan and between the Middle East

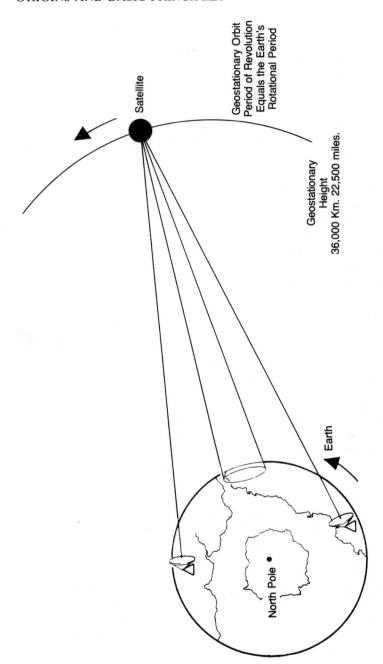

Figure 2.1   The Geostationary Orbit

and the East Coast of the USA (and which also enables the superpowers to target nuclear missiles wherever they wish). Other types of orbit are possible and these are utilised for a variety of other special purposes and applications. However, the geostationary orbit offers a number of advantages for communications purposes.

The function of a communications satellite as a "repeater in space" is analogous to that of a repeater station in a terrestrial microwave network. An example of such a network is the United Kingdom microwave network centred on the British Telecom tower in Holborn, London; the series of interlinked radio masts, in line of sight of one another, are a familiar part of the landscape. A communications satellite functions in a similar way. It receives a radio signal transmitted from earth, amplifies it to compensate for loss of power on the outward journey and retransmits it back to the receiving point or earth station.

Because of the large distances which are traversed, the signals received on earth and at the satellite are very weak indeed. For example, the signals received in the UK by the large international earth stations at Goonhilly and Madley are of the order of a few *picowatts*. (A picowatt corresponds to 10^{-12} watts or one millionth part of one millionth of a watt!) In fact, for the signals to be usable they have to be amplified several hundred thousand million times at the satellite before retransmission.

Also, in order to avoid interference between the weak incoming and the stronger outgoing signal, the frequency is changed before retransmission. The device on-board the satellite that performs the amplification and frequency conversion is called a *transponder*. A satellite carries a number of these, ranging from about 10 to as many as 50 on a large high-capacity satellite.

Until recently almost all communication satellite systems employed microwave frequencies in the 4–6 GigaHertz range, but now the 11–14 GigaHertz range is increasingly favoured and satellites are under construction or being planned to investigate the 20–30 GigaHertz range. (One GigaHertz or 1 GHz equals 10^9 Hz or cycles/second.) Frequencies in the 11–14 GigaHertz range are about ten times the frequencies used for domestic television. At these high frequencies the radio waves can be concentrated, by means of *"dish" antennae,* into beams analogous to a car's headlights. The radio frequency link through a single satellite between a transmitting earth station and a receiving earth station is referred to as the *satellite link*. It comprises an *Up-Link* and a *Down-Link* (see Figure 2.2).

In the literature the frequencies which are used on satellite links are referred to using the following convention: 6/4 GHz, 14/11 GHz, etc. The first number in each case refers to the frequency of the up-link, the second to the frequency of the down-link.

To summarise, and referring to Figure 2.2, the essential components of a satellite communications system are:

— a satellite positioned in geostationary orbit able to receive, amplify and change the frequency of a radio frequency signal before retransmitting it;

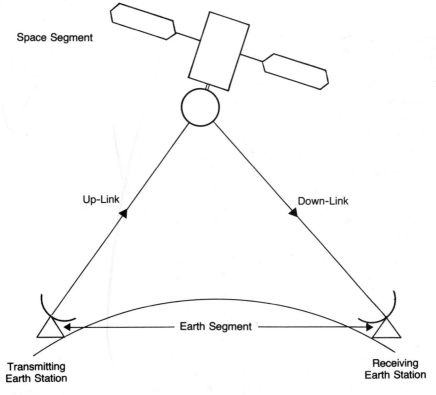

Figure 2.2   The Satellite Link

— earth stations suitably equipped to transmit and receive signals to and from nominated satellites.

The first item, comprising the satellite and the associated transmission path in space, is commonly referred to as the *space segment*. The earth station which gives access to the satellite link, and also provides the physical interface with terrestrial transmission services and end users, is termed the *earth segment*.

The Space Segment

A satellite has to operate in a particularly severe environment. Because of the practical difficulties, and cost, of intervention from earth, a high degree of automation and reliability is necessary. Only one repeater is used for a distance of 72,000 km compared with perhaps one every 50 km on earth. These conditions impose sophisticated engineering requirements, and stretch the technologies employed to the limit.

We shall now briefly review the essential components and subsystems of a satellite and the properties of the transmission path.

Satellite Construction

To carry out their relatively simple task of relaying signals, satellites require the following functional components or subsystems:

— Energy Source

For electrical power the satellite relies on solar energy which is collected by solar cells distributed either around the body of the satellite or on solar "paddles" or "sails" which are such a prominent feature of some satellite designs;

— Power Generation and Distribution

The function of this subsystem is to convert the solar energy into electrical power and to distribute this to other functional components. This also includes charging on-board batteries which provide standby power during periods when solar energy is unavailable;

— The Stabilisation Subsystem

This is to ensure that the satellite as a whole remains in a fixed orientation relative to its orbit, and that the antennae continue to point in the correct direction;

— The Thrust Subsystem

This comprises a set of on-board propulsion units and their fuel supplies. It enables adjustments to be made to the satellite's orbital position and attitude, or inclination to the orbit;

— Temperature Control System

A satellite has to endure extreme variations in temperature, and other hazards. The function of the temperature control system is to balance the temperature distribution across the exterior and within the interior of the satellite and to optimise heat dissipation;

— Transponders

The role of the transponders and their associated electronics is to receive a signal, change its frequency, amplify it and retransmit it to earth;

— The Antennae Subsystem

The purpose of the antennae – nowadays there is invariably more than one antenna on-board – is to receive the signal from space, feed the signal to the transponder, and then retransmit it.

Reception and transmission may use the same antennae, or different ones. Because of its directionality, an antenna is able to concentrate the energy into a beam and, depending upon the particular requirements, an antenna is designed so that the geographical area that the beam covers on earth has specified characteristics. The characteristics of primary interest are the shape of the illuminated area, or 'footprint', and the pattern of variation of received signal strength within the footprint;

— Telemetry, Tracking and Command System

This is quite distinct from the main operational communications system, although they usually share a transponder. Its purpose is to transmit information about the satellite to earth and to receive commands from earth;

— Communications Payload

The antennae, transponders and associated electronics are frequently referred to as the communications payload. Satellites commonly support several types of communications service, and the terminology may reflect this by distinguishing, for example, between *telecommunications payload* and *television payload*. Each type of service in this arrangement is allocated specified antennae and transponder resources.

The maximum permissible launch mass is a major constraint on the size of payload that can be carried by a specified rocket or launch vehicle, and this has a significant influence on satellite design and the trade-offs and technical innovations which are used to accommodate the payload.

The Transmission Path

The minuscule strength of the received signal is an indication of the arduous conditions the signal experiences on its journey between satellite and earth. First, much of the signal is lost to space, since an antenna is not 100% efficient. Then further losses occur due to absorption by water vapour and other forms of atmospheric precipitation. Moreover, precipitation is also a major source of noise – which distorts the signal.

Just as with terrestrial transmission services, signal loss and noise impairment are fundamental determinants of transmission link performance. Expressed in the form of the *signal-to-noise ratio* they affect both the complexity and costs of the equipment for detecting and processing the signal, and the information-carrying capacity of the link.

For these reasons the designers of satellite systems go to great lengths to maximise the signal-to-noise ratio.

The Earth Segment

Earth Stations

The earth station may be owned and operated by an organisation such as a National Telecommunications Administration, or by a private organisation. It may be accessed solely by one organisation at the site where it is located; or it may be shared between a number of users or locations. In the latter case, access from remote locations will require the use of terrestrial transmission services; and in both modes of access, arrangements for connecting to the user's internal communications services, and to ultimate end-user devices (such as telephones, terminals and computers), will also be required. Earth stations may be in fixed locations, or mobile – on ships, on road vehicles, or (soon) on aircraft.

The earliest earth stations had imposing dimensions. The dish had a diameter of perhaps 30 metres; the structure weighed about 300 tons; and the system as a whole could be rotated on a circular rail 20 metres in diameter. (The angle of elevation of the antenna dish could be varied.) The first UK earth station, erected at Goonhilly in Cornwall, provides an example.

Large dishes were required because of the weakness of the received signal (both the reception and transmission sensitivity of an antenna are proportional to the dish diameter, although other factors also play a part).

However, with the progressive increase in satellite transmission power over the years, coupled with the move towards higher transmission frequencies, it has become possible to reduce dish diameters, and to reduce the size and cost of earth stations. Nevertheless, large-dish earth stations still have a role to play, particularly for linking national telecommunications networks to international satellite services. The smaller earth stations must also have some mechanical arrangements to ensure that the antenna can be pointed accurately in the right direction, and there may also be a requirement to re-point the dish to a different satellite.

Despite the impression of sheer size and robustness, earth stations are instruments of great precision. The dishes have to be manufactured to very fine tolerance, and an antenna of 30 metres for example transmits a very narrow beam having an angle of approximately 0.1°.

Access and Utilisation

The basic earth station, comprising the transmitter, receiver electronics and antenna, merely provides access to a very-high-frequency transmission link and, via an unintelligent repeater on the satellite, a link to another earth station. For the link to be usable by ultimate end users, and for the capacity to be utilised in an efficient and cost-effective manner, there are additional requirements.

There must be arrangements for connecting an earth station to terrestrial communications infrastructures and services, and concentrating the traffic from one or more end users and organisations onto the satellite link accessed by the earth station. Secondly, just as efficient and high-quality radio communication on earth requires auditory signals to be superimposed on a higher frequency *carrier* wave, so information to be transmitted across a satellite link must in some way be impressed upon the 4 GigaHertz (or higher) satellite carrier frequency.

A transponder is an unintelligent device: it can neither distinguish between the different traffic streams (or conversations) which pass through it, nor take account of their originating or destination earth stations. All the information in the carrier is received, amplified, and its frequency changed, in a wholly transparent manner. Figure 2.3 illustrates the simplest way in which a transponder can be used to establish a link between two earth stations. If earth stations A and B are allocated specific transmit and receive frequencies, say F_A and F_A' respectively, this will provide a one-way transmission path between stations A and B, using one transponder. If, in addition, two different frequencies F_B and F_B' are allocated and a second transponder used, then a two-way communications path can be established. Therefore, in this example, a two-way link

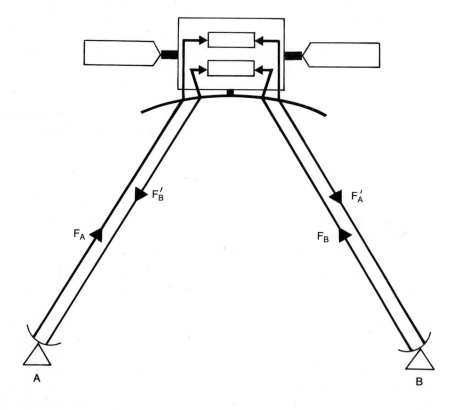

Figure 2.3 Space Segment Access by Two Earth Stations using Two Transponders

between two earth stations requires two transponders and four different frequencies. If we then extend the example by adding more earth stations and provide the capability for any earth station to communicate with any other, the transponder requirements increase dramatically.

Although satellite capacity continues to increase and costs to fall, it is apparent that this does not provide a very efficient way of utilising the capacity. Accordingly, a third requirement is to provide a mechanism which enables the satellite's high transmission capacity to be shared efficiently between earth stations and the ultimate end users.

Historically, almost all satellite services have employed schemes which only permit capacity to be allocated to multiple users in fixed amounts over preassigned routes. This can be either on a dedicated basis, so that the capacity is continuously available to the user, or in fixed amounts which are scheduled in advance. To date, the major use of satellite communications has been for applications for which this mode of access is generally suitable, and will continue to be employed for these categories of application. They include, for example: provision of national and international telephony services; the distribution of television programmes; and, in general, any application which requires either substantial permanent capacity or where the requirement may be intermittent but predictable.

However, if satellite communications is to be more widely accessible to business and the community at large and if a whole new range of application opportunities are to be exploited, more flexible and cost-effective techniques are needed. This arises because the communications requirement for many organisations, individuals and applications is not fixed, and the purchase of fixed capacity may be uneconomic.

For example, applications such as video conferencing and high-speed facsimile transmission typically require access to the required capacity for limited periods. Therefore, in addition to the fixed allocation schemes, systems have also been devised that share the satellite capacity dynamically between users according to their fluctuating demands. These various mechanisms are collectively referred to as *Multiple Access* systems, the dynamic sharing systems being distinguished from the fixed allocation schemes by the expression *Demand Assignment*. (Later in the book we shall encounter expressions such as FDMA, Frequency Division Multiple Access; TDMA, Time Division Multiple Access; and DAMA, Demand Assignment Multiple Access.)

Systems Management and Control

Within any satellite communications system certain earth stations are assigned responsibility for managing and controlling the system. In the simplest systems, this will include such tasks as: operation of the Telemetry, Tracking and Command system and monitoring the satellite's performance; and responsibility for the frequency plan and allocation of transponder capacity and frequencies to other earth stations. In more sophisticated systems, particularly those employing TDMA and Demand Assignment principles, control stations will have additional tasks to perform, and these will generally require some degree of participation and cooperation from user earth stations, so that control may become increasingly distributed rather than totally centralised.

PROPERTIES OF SATELLITE COMMUNICATIONS SYSTEMS

Satellite Communications Systems have several unique properties which distinguish them from terrestrial communications systems. The main ones are:

— the broadcast property;

— geographical flexibility;

— distance-independent cost of provision;

— high transmission capacity;

— the significant round-trip propagation delay time and the echo effect.

The Broadcast Property

That a satellite transmission can, in principle, be received at any point within its coverage area has a number of implications.

Although the manner in which an earth station receives and accesses transmitted information depends upon the particular access method employed, the procedure is roughly analogous to the operation of "tuning in" a domestic radio receiver.

The recipients in effect discard the information not intended for them and select only that which is. This means that any location, if suitably equipped, could eavesdrop on information intended for someone else with potentially serious consequences with confidential or otherwise sensitive information. Whilst the satellite system's operational and management and control procedures do present a major obstacle to unauthorised access, a determined and skilled person would stand a fair chance of gaining access to an organisation's information. Accordingly, in particular instances, it may be necessary to take appropriate defensive measures such as encrypting the information before it is transmitted.

It is not only commercial or military information that is vulnerable. Pay Television, now widespread in the US and likely to become available in the UK over the next few years, is a good example. When distributed by satellite, it is particularly vulnerable to unauthorised access – with possibly serious financial consequences for the programme companies. To prevent access by viewers who have not subscribed to the service, equipment has been developed which enables the television pictures to be transmitted in a distorted form, the TV sets belonging to bona fide customers being equipped with deciphering devices which restore the visual image. The devices themselves, together with the deciphering keys – which are altered at frequent intervals – are controlled remotely by the programme distributors.

Despite these attendant limitations the broadcast property offers great benefits. Thus, it enables the same information – whether a telephone message, a text message, a television programme or the data for printing a newspaper – to be relayed simultaneously to a number of locations, irrespective of whether or not these are served by terrestrial services.

Geographical Flexibility

Unlike terrestrial services, satellite transmission paths or networks built around them are not restricted to any particular configuration – as they would be by a predefined pattern of underground cables or microwave pylons. Within their area of coverage satellite networks offer an infinite choice of routes. An obvious advantage is that they can reach locations either lacking in or remote from terrestrial services, and this is the dominant reason why relatively undeveloped Third-World countries are so keenly interested in satellite services. In those countries where terrestrial services are either rudimentary or non-existent, satellite services may offer the most economic route to the provision of a sophisticated national telecommunications infrastructure.

This high degree of flexibility offers a number of other major benefits, even in regions with well developed terrestrial services. For example, it enables networks to be rapidly reconfigured, and links to be quickly supplied and just as readily terminated, thereby permitting temporary services which previously might have been uneconomic to provide by other means. It also gives satellite services a wide geographical "reach", enabling them to penetrate to remote locations, and hostile or physically inaccessible environments such as deserts and offshore oil platforms. It enables telecommunications services to be supplied over "thin" or low traffic routes for which terrestrial services might be grossly uneconomic. And finally, access to ships, vehicles, aircraft and other moving entities is just as easy as for fixed locations.

However, although satellite systems transcend the physical boundaries of earth, regrettably they do not escape its political or legislative pressures.

There are already strong indications that continued expansion of domestic and regional satellites and the spread of services such as direct broadcast television could become a focus for new international and political stresses. These arise mainly because satellites do not respect national geographic boundaries: although satellite beams can be shaped to conform approximately to a given geographical area, there is inevitably some "spill over" to adjacent areas or countries. (In the next chapter we encounter an early example of this in connection with the Canadian ANIK satellite.)

For some time there has been mounting concern about the possible adverse effects of invisible information flow across national frontiers via terrestrial services, and this could well intensify, with the proliferation of satellite-based services. The spread of satellite broadcast television provides an instructive and thought-provoking example of what may lie in store.

Because of its potentially wide educational and cultural implications, the international spread of broadcast television and the impending introduction of direct broadcast television is now starting to arouse misgivings in several areas of the world. This may occur, for example, because a country either prohibits advertising or has advertising standards which are different from those of the originating country; or the political leaders of a country may not wish its citizens to view programmes which promulgate uncongenial ideas.

Distance-Insensitive Costs

Whereas the cost of building and maintaining terrestrial transmission facilities is directly proportional to the length of the circuit or transmission route, this is not so for satellite services. Because the information travels through a free resource (space), one would expect the corresponding cost element in the charges made to the user, under competitive market conditions, to be largely independent of distance.

Transmission Capacity and Speeds

Transmission capacity in terms of the amount of information that the satellite is capable of handling per unit of time is determined largely by its design and by the methods employed for impressing information onto the radio frequency link. The range of transmission speeds available to the user depends upon such factors as: the types of traffic for which the system was designed; and the form in which the services are marketed.

Until a few years ago, communications-satellite technology offered a capacity and speed performance far beyond the capabilities of terrestrial transmission technology: in this respect, satellite communications could claim to have a unique advantage. This circumstance has now changed.

A few years ago, few people could have foreseen the rapid development of optical-fibre technology, and its practical implementation. The optical-fibre cables now being introduced into terrestrial cable networks have transmission capabilities comparable to or exceeding those of many of the satellites launched to date. This is easily seen by comparing the performance of an optical-fibre circuit similar to those now being introduced into the trunk telephone network, with that of a typical small-to-medium-size satellite offering small-dish business services. Such a satellite might carry ten transponders, each with a transmission capacity of perhaps 36 Megabits/second. The total transmission capacity would therefore be 360 Megabits/second. By comparison, the optical fibres now being installed in the terrestrial network transmit information at speeds of 140 Megabits/second or greater so that three of them would have a capacity in excess of that of the satellite.

Most national telecommunications administrations are committed to terrestrial network modernisation plans based upon optical-fibre technology and conversion from analogue to digital transmission. The same policy is also being followed for the transoceanic cables. Inevitably, the modernisation of the global terrestrial network will proceed in a fragmented and piecemeal fashion, and a considerable time is bound to elapse before the process is completed. Therefore, for those applications which require high-transmission speeds over continuous long-distance digital paths, the supremacy of satellite systems is unlikely to be challenged for some time.

Transmission Delay and Echo

Transmission Delay

Travelling at the speed of light, a piece of information takes about 270 milliseconds to complete the one-way trip of 44,500 miles from earth to satellite and back to earth. When allowances are also made for associated processing on-board the satellite and at the earth terminals the transmission delay is of the order of 320 milliseconds. This is quite large compared with the delays encountered on terrestrial circuits, where it is generally negligible. However, in satellite transmission the delay has a significant, although varying, impact on different classes of application and on the end users' perception of the service.

In the case of speech, a telephone user has to wait an extra 640 milliseconds to hear the reply of his respondent. For data, the major impacts are on response time for interactive types of application, and on throughput efficiency and error rates in high-volume data transmission applications. For data transmission, the effects of delay become increasingly severe with increasing transmission speed. In fact, with high-volume data transmission at high speeds over satellite links, we enter a completely different environment, as the following simple example illustrates. If we assume a transmission rate of 6 Megabits/second, and that information can be pumped continuously into the satellite channel, within the end-to-end delay time approximately two Megabits of information may be "in flight" in one direction or four Megabits in both directions simultaneously. Thus, a significant part of a user's database may be temporarily resident in space!

The Echo Effect

The echo effect, in which the speaker hears the echo of his own voice, is a feature of terrestrial telephone channels, and is caused by the presence of both two-wire and four-wire circuits and the transition between the two. Over short distances, where the transmission delay is small, its effects are negligible and hardly noticeable to the user. Over longer terrestrial distances, however, with correspondingly lengthier delays, the effects are more noticeable. For such distances, devices called *echo suppressors* are built into the network to reduce echo effects. Since a satellite telephony channel is accessed via terrestrial circuits, telephone conversations by satellite are also prone to echo. Here, because of the lengthy delay, the effects are more severe and require more sophisticated technology to overcome them.

In the context of today's terrestrial telephone networks the echo problem has been largely overcome, and acceptable solutions to the data delay problem are becoming available. But the lesson is clear: the move from terrestrial circuits to satellite links is not simply a matter of replacing one transmission service by another. The special properties of the satellite link need to be considered; and the techniques and principles which are adequate in the terrestrial domain need to be re-evaluated and modified appropriately.

3 Evolution of Satellite Communications

for guide to brief history .

INTRODUCTION

(see earlier)

The vision conjured up by Arthur C Clarke was soon to become a reality. Within twenty-five years the INTELSAT organisation had stationed satellites over the Atlantic, Pacific and Indian Oceans to form a global network providing communications coverage of almost the entire earth. In its relatively brief history, communications-satellite technology has experienced a dramatic growth, each generation of satellites following another in rapid succession and incorporating major advances.

In this chapter we highlight events of major significance, chart the progress in satellite design and satellite capabilities, and chronicle the changing perceptions of applications in this field. The chief characteristics of the various satellites that are considered are listed in Tables 3.1 to 3.4 (see pages 61 to 66).

THE EXPERIMENTAL PHASE

In October 1957, the Earth was presented with its first artificial satellite – SPUTNIK I (Table 3.1), and the age of space communications dawned with the repetitive "bleep-bleep" signal received from this device. The USSR quickly followed this success with SPUTNIK II which carried the first living passenger – a dog! – to be transported by a space vehicle. Following SPUTNIK, the USA launched EXPLORER I which made a number of important scientific discoveries. However, none of these satellites was designed specifically for communication purposes.

The world's first active communications satellite was SCORE, launched by the US National Aeronautics and Space Administration (NASA). It was referred to as an *active* satellite because it had an energy

41

source on board, and was able to broadcast a message pre-recorded on magnetic tape; this was President Eisenhower's Christmas message to the nation. Because the message was pre-recorded it did not function as a true repeater. This satellite only survived for twelve days.

In the early 1960s, there were also US experiments with large metal-coated balloons. These had diameters of about 30 metres and were inflated in orbit at altitudes exceeding 1,000 km. Their purpose was to reflect radio energy, as does radar. These did not carry any energy sources on board and did not act as repeaters.

These devices and other satellites which do not possess a transmission capability are sometimes referred to as "passive" satellites. Despite the aluminium coating, quite high transmission power (10 kws) was required for the signals from earth to obtain an adequate return echo. ECHO I and II are examples of this class of satellite.

Probably the world's first fully active repeater satellite was COURIER, launched by the US Defence Department. This successfully transmitted speech and telegraphy, but it also had a short life of seventeen days. It was the first satellite to use solar-powered cells.

The era of satellite communications opened properly with TELSTAR I in July 1962. This device was launched by the American Telephone and Telegraph Co (AT&T) and was the first active satellite carrying on-board antennae and amplification equipment. It was placed in elliptical orbit with an orbit transit time of 2 hours 40 minutes, and was simultaneously visible from two earth stations for only about ten minutes. TELSTAR was spherical, with a diameter of 86 cms and a mass of 80kg.

Transmission frequencies in the 6/4 GHz range were used, and the single 2-watt transponder had a capacity equivalent to 600 telephone circuits. It had a life span of one year achieved through the combined use of solar cells and rechargeable batteries. TELSTAR I and the later TELSTAR II provided the first intercontinental links between the USA and Europe transmitting both telephony and live television. International live television broadcasting was a very important early application for satellites, since transmission of television signals was beyond the capability of the existing transatlantic cables. The role of the TELSTAR satellites was subsequently taken over by RELAY I and RELAY II. These had the same capacity but the length of time during which they were visible increased to thirty minutes.

THE FIRST GEOSTATIONARY SATELLITE

All the satellites mentioned above were stationed at quite low altitudes in mainly elliptical orbits. Therefore they did not appear stationary relative to earth as they would in geostationary orbit. This was largely because the rockets then available could not boost satellites into orbits higher than 10,000 km above the earth. In consequence, the earth stations had to be steerable in order to track the satellite. An additional shortcoming was that the satellite and the earth station did not remain in line of sight contact for very long. Thus, not only was TELSTAR only visible for about thirty minutes, but an earth station antenna such as the 1100 ton array at Goonhilly in Cornwall had to be nimble enough to follow it during that time.

However, in 1964 the USA placed SYNCOM 3 in a stationary orbit, and this very quickly demonstrated the advantages of the geostationary orbit, thus sounding the death knell for low orbiting systems. A noteworthy achievement of SYNCOM was the first continuous live trans-pacific television broadcast of the 1964 Olympic Games in Japan.

Since SYNCOM, the number of geostationary systems has continued to increase. A notable exception is the Soviet MOLNYA, which was launched in 1965. MOLNYA describes a low-altitude, highly-elongated elliptical orbit, and is currently the only civil telecommunications satellite employing a low altitude orbit. However, there are sound geographical and economic reasons for using such an orbit (see Chapter 4).

At an early stage of development, the USSR took the decision to go for high satellite transmission power (40 watts/transponder). This made it possible to use far smaller earth stations than those of the INTELSAT organisation. In fact the international part of Soviet communications is very small and the MOLNYA satellites primarily serve domestic requirements.

EXPANSION AND GROWTH

TELSTAR, RELAY and SYNCOM established the feasibility of satellite communications, and marked the beginning of the commercial satellite communications era. The world had embarked upon a process of expansion and growth that is still in progress.

The enthusiasm and excitement which greeted the arrival of satellite communications was by no means universal. On the one hand, there were vested interests – in the shape of the various telecommunication administrations, anxious to protect the revenue-earning potential of their vast investments in terrestrial systems; and on the other hand, there was the framework of regulations which govern the operation and supply of telecommunications services in many countries and also internationally. Thus, the rate of introduction and exploitation of satellite services has been constrained almost as much by regulatory or political factors as by technological ones.

Because of the central role of telecommunications in shaping national economies, social structures and the relations of countries with one another, it has been subjected to some form of governmental control (or *regulation*) in all advanced nations. The regulatory framework employed within a country and internationally can exert a significant influence on the nature, supply and utilisation of telecommunication services.

TELSTAR affords a good example of how the regulatory environment can affect the course of events. Bell Laboratories, in the studies that yielded TELSTAR, showed that a few powerful satellites of advanced design could handle far more traffic than the entire AT&T long-distance networks, and that the cost of the satellites would have been a fraction of the cost of the equivalent terrestrial facilities. However, the existing regulations, laid down by the United States Federal Communications Commission (FCC), prohibited AT&T from entering satellite communications.

The field was therefore opened up to competition, and a number of corporations, minute in size compared with AT&T, announced their intention of offering satellite communication services, whilst AT&T continued to invest millions of dollars in its terrestrial networks. It should therefore be borne in mind that the development of satellite communications has been partly dependent upon successive changes in the regulatory environments of countries, particularly the USA, and, in the international sphere, through the establishment of international agreements.

Whilst this is not the place to explore the complexities of either the American or the international regulatory scene, the following key events are worth noting:

1962   The US Federal Communications Satellite Act.

1962   Formation of the Communications Satellite Corporation (COMSAT).

1971   Federal Common Carrier Decision.

1972   Federal "Open Skies" policy.

1982   Federal interim rules governing the operation of Television Direct Broadcast Satellite Services.

The Federal Communications Satellite Act has a special significance, because it was the first attempt to define and regulate the operation of satellite services in the US. The most important consequence of the Act was the formation of the Communications Satellite Corporation (COMSAT) which stipulated, among other things, that only COMSAT could operate and supply satellite services in the US, and, furthermore, that the services could only be used for international transmission. The other regulatory decisions are discussed below.

There is a discernible pattern to the application and exploitation of communications satellites. During the first decade, their sole use – apart from military and scientific applications – was to provide international links between the national telecommunications networks of the participating countries.

Moving into the second decade, attention shifted towards satellite services having a more restricted coverage – either of a single country or a geographical region. The introduction of domestic satellites also triggered off in the US a largely unforeseen growth in television broadcasting by satellite, the programmes being received by community antennae (CATV) for onward distribution by cable or microwave systems. Since the late 1970s other types of application have been investigated and are progressively being introduced. In particular, the success of the maritime satellite communications system operated by the INMARSAT organisation established the feasibility of communications between earth stations which were essentially mobile rather than fixed.

From the time of their introduction, satellite services, whether international or regional, have carried business and commercial traffic, including telephony, data, telex and facsimile. However, the form in which the services have been accessed and supplied has not been sufficiently flexible or economic to meet specialised business requirements, and this has tended to discourage the more widespread exploitation by business, particularly for the more novel types of application. As far as international transmission was concerned individual users had no choice, since access was controlled by the national PTT or Common Carrier. On the other hand, where direct access was available – on some of the US domestic services, for instance – the high earth station costs discouraged all but the large corporations. These limitations have provided a powerful stimulus to the development of services geared to the needs of the business community. This is expected to be a major growth area during the next decade.

The next major service development, now in its formative stage, is Direct Broadcasting by Satellite. Technical progress over the last few years has now made it possible to construct satellites which can broadcast TV programs direct to home antennae at relatively low cost, thus bypassing existing cable or terrestrial microwave distribution networks.

INTERNATIONAL SERVICES

As satellite costs dropped from their initially high levels, it was quickly realised that they could compete with the world's suboceanic cables. Although there was some initial resistance by the cable owners to the use of satellites, they and the National Telecommunications Administrations quickly appreciated the opportunities they presented.

During the 1950s, cables and radio links provided relatively satisfactory communications within countries and continents. In contrast, long intercontinental links or links across deserts and similar regions were infrequent and very costly. In 1956 the first transatlantic telephone cable (TAT–1) came into operation. However, although this could handle 56 simultaneous telephone conversations, it soon became apparent that submarine cables alone could not handle the projected growth in telephone traffic. (In 1982 in the United Kingdom alone, the volume of international telephone calls was still growing at an average annual rate of 20%, and showed no signs of slackening.) Nor could much assistance be expected from terrestrial radio, because this was not only of indifferent quality due to atmospheric interference, but the relevant portion of the radio frequency spectrum was becoming overcrowded.

In 1964 a number of countries signed an agreement to establish a worldwide satellite telecommunications service. This was the beginning of the INTELSAT organisation, although the body was not finally constituted until 1971. INTELSAT holds the monopoly for the provision of *international* public and private transmission services over satellite links; it also leases satellite capacity to a number of countries for their domestic use, and a growing number of national and regional services also lease capacity to INTELSAT.

There are now mounting pressures for greater competition in international telecommunications, and if these succeed, then INTELSAT's monopoly could be eroded. Nevertheless, whatever one's view on its monopoly position, by any standards INTELSAT is a remarkable example of international cooperation. Not only has it played such a dominant role in establishing global satellite communications services but it has also made very substantial contributions to technological development. The following statistics illustrate the global importance of INTELSAT:

INTELSAT Statistics (1982)
Member countries 106
Satellites in orbit 16
Operational Earth Stations 216, at 156 locations in 136 countries

Traffic statistics

60% of traffic on intercontinental routes
26,000 full-time voice and data channels
20,000 scheduled television transmissions/annum
Annual traffic growth rate: over 20%

Since it commenced activities, INTELSAT has commissioned six classes of satellite, starting with INTELSAT I; and the latest, INTELSAT VI, is planned to become operational in 1986. In each class a number of models have been produced and launched.

The INTELSAT series constitutes the longest continuous record of development in satellite communications generally and can be taken as representative of the progress of satellite-communications technology as a whole. For that reason it is useful to review the main characteristics of the various models (listed in Table 3.2).

INTELSAT SERIES

INTELSAT I (Early Bird)

On the 6th April 1965, the small but now legendary Early Bird satellite was brought on station over the Atlantic; for the first time, it was possible to establish commercial intercontinental telephone links between Europe and North America. It handled 240 telephone channels or alternatively one TV channel. However, it had two major limitations: it was necessary to interrupt the telephone service during TV transmission because it had insufficient power for both; and it could only work with two earth stations at a time. Because of the latter restriction, a rotation arrangement was operated, earth stations in different countries being allocated one week of activity in turn.

Nevertheless, Early Bird demonstrated the viability of commercial satellite services, and also proved that communications satellites could be extremely reliable. Built for a nominal life time of eighteen months, it was only taken out of service in 1970, more than five years after launch.

Like all satellites in the series, Early Bird was spin stabilised. (In order to ensure that the cylindrical satellite maintains a constant orientation with respect to earth, it is continuously spun about its vertical axis.) It was also equipped with a non-directional antenna that transmitted radiation in all directions, so that a major part of the radiated power was lost to space.

INTELSAT II

The several versions of INTELSAT II that were launched had much in common with their predecessor, but benefited from some major improvements. Capacity was still 240 telephone channels or one TV channel, together with spin stabilisation and a non-directional antenna. However, transmitter power was considerably increased (from six watts to eighteen watts) through the use of a new transponder design. The principal innovation lay in the introduction of a multiple-access technique which enabled all connected earth stations to converse simultaneously. This facility became available from 1967 onwards with three satellites in orbit (one over the Atlantic and two over the Pacific).

INTELSAT III

The first INTELSAT III made its appearance in 1968. The major innovation was the incorporation of a contra-rotating directional antenna. This device concentrated the power in the direction of earth; and by rotating the antenna assembly in a direction opposite to that of the satellite body, the pointing direction remained fixed relative to earth.

As Table 3.2 indicates, this resulted in a significant increase in performance. The INTELSAT III satellites were positioned over the three main oceanic regions – the Atlantic, Pacific and Indian oceans – and by 1969 they were providing global coverage for the first time.

INTELSAT IV

The eight satellites in the INTELSAT IV series were also characterised by a sharp increase in capacity. This was achieved through increasing the number of transponders (12) and adopting a new antennae configuration. In addition to an earth coverage antenna this had four reflectors, providing hemispherical coverage and two spot beams. The direction of the spot beam antennae could be changed by remote control from earth, making it possible to direct high-capacity beams to small areas, depending upon demand and traffic density.

INTELSAT IVA

This was a series of eight satellites developed from INTELSAT IV. A major innovation was a complex antennae configuration which separated the spot beams serving the eastern and western hemispheres. This enabled the 6/4 GHz frequency band to be used twice over without mutual interference. Thus with no significant increase in satellite payload it was possible to increase capacity without utilising new frequency bands, or employing a duplicate operational satellite.

INTELSAT V

INTELSAT V differs radically from the previous series. Whereas the latter used spin stabilisation, this was not practicable with INTELSAT V, which is cubic rather than drum shaped. Instead it is stabilised along each of its three axes using complex automatic control systems. Like its predecessor, it uses separated beams to achieve frequency re-use, but in addition polarisation is also employed, thereby achieving four-fold re-use of the frequency spectrum. Also the satellite operates in the 14/11 GHz range in addition to the 6/4 GHz range. These innovations have made it possible to double the telephone channel capacity compared with INTELSAT IVA.

Later versions of INTELSAT V scheduled for 1984 will incorporate further modifications and will provide a harmonious transition to INTELSAT VI scheduled for launch in 1986.

INTELSAT VI

This series, whilst reverting to the spin stabilisation of its predecessors, will incorporate some major innovations. First, it will offer six-fold frequency re-use, achieved partly through the techniques mentioned earlier and also by employing digital transmission. It will also incorporate onboard switching equipment which will enable the traffic received from an earth station on one beam to be instantly switched to an earth station connected to another beam. In effect, this converts the satellite into a telephone exchange in space.

The first half dozen satellites in the series are planned to have an individual capacity of 40,000 telephone channels. In respect of its physical size, and its advanced technical features and capabilities, INTELSAT VI can be taken as typical of the class of large, high-capacity satellites which are being planned or constructed for the mid-1980s onwards.

As far as the INTELSAT network as a whole is concerned, several developments are under way. Prior to the launch of INTELSAT VI, a start is being made on conversion to digital transmission and the introduction of the more efficient Time Division Multiple Access technique. Although its major role will continue to be the provision of global telephony and associated services between countries, plans are being formulated to offer small-dish specialised business services using the 14/11 GHz frequency band.

EMERGENCE OF NATIONAL AND REGIONAL SERVICES

When the first INTELSAT satellites brought competition to suboceanic cables, the US domestic telephone networks seemed immune from the threat. The cost per telephone channel of the early satellites was still high. In addition, the US Communications Satellite Act had established COMSAT as the sole operator of satellite services and, furthermore, restricted their use to international communications.

The first satellite to be launched primarily for national purposes was the Russian MOLNYA. But Canada, because of its widely-separated population and often harsh and inhospitable environment, was the first non-Soviet country to recognise the potential of a geostationary satellite system for extending the benefits of advanced communications services to isolated communities, and encouraging human settlement and economic expansion. Canada established TELESAT Canada in 1969 and launched the ANIK A satellite in 1972 (ANIK means "brother" in Eskimo). Since then a number of countries have followed suit: these include the USA, Indonesia and Mexico. Others are planned by countries such as the UK, France, Australia, India, Brazil, and Colombia.

At the time when ANIK became operational in Canada, technology was changing in the US more rapidly than the law. The area covered by ANIK "spilled over" into US territory, and some American organisations were quick to set up antennae to access ANIK, although this was of dubious legality. However, in 1971 the FCC (Federal Common Carrier) decision had effectively opened up to competition the terrestrial transmission market in which AT&T had hitherto enjoyed a virtual monopoly. Therefore in 1972, in order to regularise the position regarding domestic satellites, the Federal Government promulgated the "Open Skies Policy" which encouraged private industry to submit proposals for launching and operating communications satellites. This led to Western Union's launching of WESTAR in 1974, RCA's launching of SATCOM in 1975, and AT&T's launching of COMSTAR in 1976.

The use of satellite communications in the State of Alaska is a fascinating, although little publicised, example, which deserves mention. It is of interest, not only because the nature of the terrain, with its isolated communities, provides an ideal opportunity, but also because of its early pioneering of innovative applications like teleconferencing and educational broadcasting.

A major stimulus came from the North Slope oilfield and pipeline developments, and experiments started in the early 1970s using the ATS–1 and ATS–2 satellites. As a result of cooperation between the State, RCA and Alaska Communications (ALASCOM), Alaska now has what must still be regarded as a remarkably sophisticated communications infrastructure.

The characteristics of a selection of these national satellites are presented in Table 3.3.

Countries belonging to the same geographical zone may wish to strengthen their communications resources, and a number of regional satellite services are in the process of being established, a few of which are listed in Table 3.4. A regional service is distinguished not only by its territorial coverage but also by the constraints and technical options imposed by the plurality of countries, regulations and languages concerned.

Although satellite communication services have been under consideration within Europe for some time, they have been much slower to develop
than in the USA. Although the distances across Europe are not greatly
different from those across the USA, such factors as the diversity of
languages and cultures, different telecommunications administrations
and regulatory regimes, and a lower level of potential customer demand
have inhibited the growth of satellite communications. Furthermore,
whilst a requirement could be seen for a European regional satellite
service, it was less easy to justify the case for exclusive national systems.
Therefore, since 1975 the principal countries in Europe have co-
ordinated their space research and satellite communications development programmes under the aegis of the European Space Agency (ESA).
The EUTELSAT organisation has been established to commission the
design and construction of satellites and to manage the operation of
European-wide services on behalf of the participating countries.

The first communications satellite to be launched was the Orbital Test
Satellite (OTS) in 1978, and this is still in operation three years beyond its
planned life span. Apart from clearly demonstrating a European capability, this satellite has been successfully applied to a variety of experimental
and operational purposes. These have included: experiments with
small-dish antennae in the 12.5 to 12.7 GHz range; experiments with
video-conferencing, high-speed facsimile transmission, news distribution
and printing; investigations into the interconnection of local area networks in the Project Universe Experiment; and many others besides.

The ECS–1 satellite successfully launched on the 15th June 1983 is the
first in the series of communications satellites which are planned to come
into operation over the next decade. The system will provide an intra-
European telephone system for the member countries of the European
Conference of Posts and Telecommunications, and there is provision for
the exchange of TV programmes between members of the European
Broadcasting Union.

Initially, access will be obtained through a limited number of national
earth stations with 15/19 metre antennae, built to the INTELSAT Earth
Station C standard. Commencing with the next satellite in the series
(ECS–2) provision will be made for business services which can be
accessed through earth-station antennae of five metres or less.

It is planned to launch, from 1986 onwards, a series of considerably more advanced satellites. The first of these is referred to as L–SAT, although the consortium headed by British Aerospace, which will be building the satellite, is marketing it more widely as a member of a family of satellites designated the OLYMPUS series.

L–SAT involves state-of-the-art technology; apart from its operational use, it will be used as a test bed for several advanced facilities and concepts; for instance, it will be used to investigate on-board satellite switching of traffic streams and transmission performance in the 20 to 30 GHz range.

The example of Europe has been followed by a number of international organisations and country groupings wishing to create systems which will encourage economic development and tighten the links between members. Thus, countries of the Arab League have come together to form the ARABSAT organisation for this purpose. Similarly, countries in South East Asia are joining with Indonesia for the purpose of extending the Indonesian PALAPA system on a regional basis. Other projects are under study in South America amongst the Andean countries, and in Africa.

THE IMPACT OF SATELLITE BROADCAST TELEVISION

In the early 1970s, a trend of major significance emerged in the US. This was the largely unforeseen growth of satellite communications to broadcast television pictures to Community Antennae, for onward distribution to homes by cable (CATV). Whilst common carriers and corporate users saw the satellite as providing two-way channels between relatively few earth stations, the broadcasting industry perceived it as a potentially ideal way to distribute one-way signals, enabling television or radio programmes to be broadcast over vast areas. This area of application developed at a very rapid rate, and had the effect of injecting a new vitality into the US CATV industry and into programme production and film companies. It also provided a strong financial base for the further development and growth of satellite services.

It now seems to be generally acknowledged that most US domestic satellite services would have difficulty in maintaining financial viability if they had to rely entirely on telephone and data traffic. Currently, almost all of the programming is aimed at entertainment, and the educational and wider cultural opportunities have not yet been explored. To further illuminate the importance of TV distribution in the US context, a few statistics are worth quoting:

> 70% of the approximately 80 million US households are within reach of cable.

> About two thirds of television sets now receive their input by satellite via local cable or terrestrial microwave distribution systems.

> In 1980 one TV programming company alone (WTBS) had about 9 million subscribers spread across 2,200 cable systems. And in the same year the SATCOM–1 satellite supplied via its earth stations about 2,900 cable systems serving 12 million homes.

MOBILE SATELLITE COMMUNICATIONS

Until the introduction of maritime satellite communications systems to serve ships at sea, all satellite communication links were between fixed locations. The first maritime system (MARISAT) was established in the mid-1970s by an American consortium. Shortly afterwards Europe also decided to establish such a network, but this was quickly superseded by an international agreement to establish a global system. The programme is managed and co-ordinated by the INMARSAT organisation with headquarters in London. The system comprises three satellites MARECS–A MARECS–B, and MARECS–C. MARECS–A is established over the Atlantic and when the others are positioned, they will provide complete coverage of the Atlantic, Pacific and Indian oceans.

The INMARSAT system provides telephone and telex links between ships at sea and coastal stations which can relay ship-to-shore search and rescue messages, in emergencies. Each satellite can handle up to 40 high-quality telephone circuits. The services provided by INMARSAT extend beyond these obvious ones, and include communications to oil rigs; data relating to ship performance, routeing and cargo manifests; and the provision of very accurate navigational data.

To initiate a distress signal, a special key on board the ship effects an immediate connection to the most appropriately located earth station via priority access to the satellite space segment. The coastal earth stations in Europe are currently located at: Eik (Norway), Goonhilly (England), Odena (Russia), and Fucino (Italy), although many more installations are planned, or under construction.

The introduction of maritime services has served to focus attention on the whole field of mobile communications and opened up a range of novel opportunities. If earth stations could be made small enough and economically priced they could be carried not just on ships, but on the backs of lorries and other vehicles, on aircraft, and conceivably in the longer term by people. And they could also be transported and sited at business and other premises to meet short-term communication needs.

THE EMERGENCE OF SMALL-DISH SYSTEMS

The characteristics of the INTELSAT system and most of the other systems we have described were largely dictated by the low power of the earlier satellites and the almost universal use of the 6/4 GHz frequency band. These features were responsible for a number of limitations:

— the choice of frequency combined with low power necessitated expensive earth stations with large antennae reflectors (anything between 45–100 ft in diameter);

— satellites share the 6/4 GHz frequency band with terrestrial microwave systems, and the respective transmissions can interfere with one another;

— mainly because of the interference problem, but also for physical and possibly environmental reasons, earth stations had to be sited in areas remote from business and residential areas. Thus, it was not possible to locate these substantial engineering structures in a parking lot, or on the roof of an office building or a house;

— as a result access to an earth station generally entailed the use of possibly lengthy and costly terrestrial circuits, and this could make it prohibitively expensive for many users and their applications;

— apart from these considerations a cause for growing concern was the increasing demand for geostationary orbit space and the projected overcrowding of the geostationary orbit.

These and a number of related issues were starting to be addressed by the mid-1970s. It was already known on theoretical grounds that, by increasing satellite transmission power and moving up to higher transmission frequencies such as 14/11 GHz and beyond, it was possible to concentrate transmission power into quite narrow beams that could be received by relatively unsophisticated earth stations with small diameter antennae (5 metres or less). A few satellites with these capabilities started to come into service about this time, although mainly for experimental purposes.

This new generation of satellites combining high power with smaller lower-cost earth stations is opening up a new era in satellite communications. Two areas where its impact is already becoming apparent is in Direct Broadcast Television by Satellite (DBS) and the provision of specialised business services.

DIRECT BROADCAST TELEVISION

The proposed Direct Broadcast Television satellites are designed to transmit high-power signals which can be received by relatively inexpensive equipment. The high signal strength combined with small antenna dimensions accounts for part of the cost reduction, but additional major cost savings arise because for simple broadcast reception the equipment need only have a receiving capability.

These Television Receive Only (TVRO) earth stations, mounted on roof tops, parking lots, and in back gardens, will enable homes and offices to receive a wide range of television programmes and information services. In addition to direct reception the DBS signals can also be used to feed community earth stations for onwards distribution to cable and other terrestrial local distribution systems, as they are at present.

Some early experiments with DBS had been carried out throughout the seventies, and in the US there were already a relatively small number of consumers able to afford expensive private earth stations. These experiments utilised the conventional low-power satellites which transmitted TV broadcast signals with between 5 to 20 watts of power and earth station antennae typically 15-20ft in diameter. The high power of the DBS satellite is the real key, transmitting signals at 200 watts which can be received by antennae measuring 2½ feet or less in diameter. The technology is now in place which will enable the DBS satellites expected to come into service in 1986 to transmit to antennae with diameters of the order of 1.5 feet.

In 1982 the FCC gave the go-ahead for the introduction of DBS and invited applications for licences. Within a short time, eight organisations applied and most of these have been successful. Although there are plans for DBS services in Europe, attitudes differ in different countries, and some countries have not yet formulated firm policies. However, in the UK, it is planned to provide two DBS channels on the UK UNISAT domestic satellite due to be launched in 1986, and France and Germany have a joint agreement to launch experimental satellites: TV–SAT (Germany) and TDF1 (France).

SPECIALISED BUSINESS SERVICES

Business and commercial organisations have for some time been using satellite transmission: in the US, organisations have had direct access to domestic satellite services, and a substantial volume of international telephone and data traffic is carried by the INTELSAT system.

However, for several reasons the majority of existing services are not well adapted to business requirements, so that the business community has been unable to fully explore and utilise the potentialities of satellite communications, particularly for those applications not currently feasible using terrestrial services.

The fact is that the characteristics of the majority of satellite communications systems have continued to reflect their first major application, namely the transmission of telephony traffic and more recently TV distribution. For example, a satellite's transponder capacity was invariably subdivided into basic speech channels of fixed bandwidth, and prospective customers have commonly been committed to leasing capacity on a dedicated basis, irrespective of their actual requirements, and in addition the minimum capacity that could be leased might also exceed total requirements. Thus, although an organisation interested in investigating the benefits of video teleconferencing might be able to lease the 2 Mbits/second channel necessary for this application, this might have had to be leased for 24 hours a day for an extended period.

Thus, the earlier service offerings were not well matched to the requirements of business nor designed to encourage the wider exploitation of satellite communications for quite novel applications. For many applications which necessitate or justify high transmission speeds, users typically require access on-demand for short time durations.

A communications satellite link is much more than a point-to-point circuit or a "cable in the sky": a signal sent up to the satellite reaches every point within its area of coverage. At the same time the satellite itself is an expensive resource, and the geostationary orbit has a finite capacity,

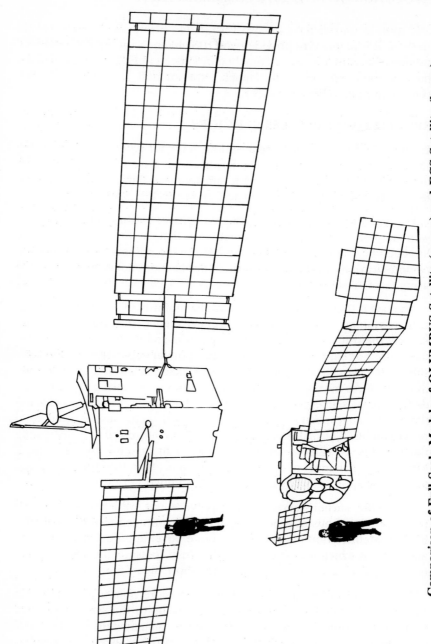

Comparison of Full Scale Models of OLYMPUS Satellite (upper) and ECS Satellite (lower)

both very good reasons for ensuring that it is efficiently utilised. It was realised that in order to achieve better utilisations and to extend its use to a wider population of end users, the answer lay in sharing the capacity between many users, and making capacity available on-demand to each of the users, according to their individual needs.

Thus, any user should be able to request a portion of the capacity at any time and have it allocated to him as soon as capacity is available. When he has finished with it the capacity would be relinquished for use by other users. Over the last few years considerable efforts have been devoted to achieving this *multiple-access* capability, and ingenious ways have been devised of allocating satellite capacity to geographically scattered users, permitting them to intercommunicate. The more sophisticated and efficient systems require the use of *digital* rather than *analogue* transmission.

The first service with these characteristics and aimed specifically at the business community was conceived by Satellite Business Systems (SBS) in 1977. Following field trials the first satellite was launched in 1980; and a fully operational service was opened in 1981.

TECHNOLOGICAL TRENDS IN SATELLITE DESIGN

The rate of technological advance which has enabled satellite communications to reach its present state continues unabated, and will continue throughout the decades ahead.

The historic trend towards larger and higher capacity satellites, which is well illustrated by the INTELSAT series, will continue. For example, in a little over 15 years the payload has risen from 38.5 kg (Early Bird) to 1,012 kg (INTELSAT V), and a mass of about 2,000 kg is planned for INTELSAT VI. Early Bird is minuscule in comparison, being a cylinder about 28″ high and 23″ in diameter, whereas for INTELSAT VI the corresponding figures are roughly 18′ and 10′ respectively. The figure on the facing page, comparing the ECS and OLYMPUS satellites, further emphasises this dramatic growth.

Economy of scale effects continue to be of central importance in the economics of satellite communications. The increases in power, capacity and overall cost/performance improvement which result from increasing satellite size and payload may easily outweigh the associated weight and cost penalties. Other scale benefits derive from improving the efficiency of utilisation and maximising the sharing of capacity. These considerations will continue to exert a major influence on satellite communications technology.

The investigation and utilisation of higher transmission frequencies, which started with the move from the 6/4 GHz range to 14/11 GHz, will continue. Experiments in the 30/20 GHz range are planned and the National Aeronautics and Space Administration has been investigating the feasibility of using 80 GHz and higher. These higher frequencies, apart from alleviating the pressures on the increasingly crowded lower end of the frequency spectrum, also hold out the promise of even further reductions in earth station antennae size.

In common with the growing adoption of digital transmission principles in terrestrial networks, satellite communications is also undergoing a similar transformation. The very great benefits arising from digital technology are becoming available on satellite channels, the important difference being that for satellite communication, the benefits will become available over long distances and on a global scale much earlier.

There are very important trade-offs in satellite system design between the space segment and earth segment costs. This has already been implied in some of the trends which have been noted: a larger satellite with increased transmission capacity, and larger and multiple spot beam antennae costs more, but it results in cost savings on the ground. This relationship is likely to be increasingly exploited in future generations of satellites.

Further developments which are planned involve taking advantage of microprocessor technology and placing more intelligence on board the satellite. Of particular interest is the provision of on-board switching capability. So far, in this book, the satellite has been regarded as a device which merely receives and retransmits information, the actual assignment of channels to users and the geographical routes being determined and controlled from the ground. This has also applied to those satellites which have more than one antennae and which direct multiple beams to different parts of the earth. The actual routes between users connected to the different beams is generally pre-assigned and is determined by the user's route requirements and the overall traffic patterns.

With the objectives of providing yet greater flexibility of use and improved utilisation, studies have been under way for some time of satellites with an on-board switching capability.

Employed in conjunction with multiple "spot" beams illuminating different parts of the earth, the satellite switch would rapidly change the connections between the spot beam antennae, so that each beam is connected in sequence to each of the other antennae beams; thus each earth station is allowed to communicate with each other station. In effect this converts the satellite into the space equivalent of a telephone exchange. In addition to providing a greatly enhanced inter-communication flexibility for users, it will also enable the satellite system to follow variations in demand and to match demand to capacity much more efficiently. Several satellites with this capability (eg INTELSAT VI and the European L–SAT) are planned or under construction.

A further refinement for the more distant future is to equip a satellite with a scanning beam that "polls" the earth stations in turn, inviting each of them to receive or transmit during the brief period of contact.

Notes on Tables 3.1 to 3.4
Communications Satellite Characteristics

1. The tables list the main characteristics (where these are known) of a number of operational or planned communications satellites.

2. The information has been gathered from a number of sources and, as far as practicable, verified by cross-reference. However, some minor residual inaccuracies may still remain.

3. The mass which is quoted is the in orbit mass, generally at the start of the satellite's life. The mass at launch may be twice this figure.

4. When two heights are specified, this indicates that the orbit is elliptical, with the earth at one focus of the ellipse. The first or smallest distance denotes the *perigee* when the satellite is closest to the earth, and the second or largest distance denotes the *apogee* when it is farthest from the earth.

5. The power which is quoted is the power of an individual transponder. It does not equate to the power actually radiated by the satellite antenna (EIRP), nor the power received on the ground.

| Name Sponsor/launch authority | Launch Date | Orbit Mass(kg) | Transponders | | | Orbit (Kilometres) | Life | Capacity | Additional Comments |
			No.	Frequency (GHz)	Power (Watts)				
SPUTNIK 1 (USSR)	1957	—	—	—	—	227 946	—	—	First artificial space satellite
SPUTNIK 2 (USSR)	1957	—	—	—	—	225 1,671	—	—	First live passenger – a dog!
EXPLORER 1 (USA)	1958	30	—	—	—	360 2,550	—	—	Discovered Van Allen Radiation Belt
SCORE (USA: NASA)	1958	150	—	—	8	180 1,490	12 days	—	Probably world's first active satellite Broadcast pre-recorded tape Battery Power only, therefore limited life
ECHO I (USA: NASA)	1960	60	—	2.5/1	Passive	1,600	—	—	Circular orbit
COURIER (US MILITARY)	1960	500	—	1.9/1.7	4	966 1,204	17 days	—	Probably world's first fully active repeater satellite Used solar power cells
TELSTAR 1 (AT&T)	1962	80	1	6/4	2	950 9,650	1 year	Telephony channels 600	First to transmit and receive simultaneously Used solar cells and rechargeable batteries
RELAY 1 (RCA)	1962	80	1	4/1.7	10	1,300 7,500	—	Telephony channels 600	
SYNCOM 2 (Hughes Aircraft)	1963	40	2	7.4/1.8	2	Geostationary	—	Telephony channels 300	First geostationary communications satellite Relied entirely on solar cells
MOLNYA (USSR)	1965	1000	3	1/.8	40	500 40,000	—	Telephony 240 (per transponder)	

Table 3.1   Communications Satellites : Early History

Name Sponsor/launch authority	Launch Date	Mass(kg)	Transponders			Orbit	Life	Capacity		Additional Comments
			No.	Frequency (GHz)	Power (Watts)			Telephony	TV	
INTELSAT I (EARLY BIRD)	1965	38	2	6/4	6	Geostationary	18 months	240 or	1	
INTELSAT II	1966	87	1	6/4	18	Geostationary	3 years	240 or	1	
INTELSAT III	1968	146	2	6/4	11	Geostationary	5 years	1,500 plus	2	
INTELSAT IV	1971	709	12	6/4	6	Geostationary	7 years	4,000 plus	2	
INTELSAT IVA	1975	862	20	6/4	8	Geostationary	7 years	6,000 plus	2	
INTELSAT V	1980	1,102	21 6	6/4 14/11	8 20	Geostationary	7 years	12,000 plus	2	
INTELSAT VI	1986	2,000	50	6/4 14/11	5.5–16 8.5	Geostationary	10 years	up to 40,000 plus	4	

Table 3.2   Communications Satellites : The INTELSAT Series

Name Sponsor/launch authority	Launch Date	Orbit Mass (kg)	Transponders					Life	Capacity	Comments
			No.	Frequency (GHz)	Bandwidth (MHz)	Power (Watts)	EIRP (dbw)			
ANIK (A,B,C,D series) (Canada)										
ANIK A	1972 onwards	—	12	6/4	36	—	33	7	6,000 telephony *or* 12 colour TV	Provide telephony, TV, associated services to remote areas Teleconferencing, education/health care
ANIK D	1982 onwards	—	24	6/4	34	—	36	7		
WESTAR (USA: Western Union)	1972 1975 1979	—	12	6/4	36	—	33	7	7,200 telephone channels *or* 12 colour TV	First US domestic satellite Handles services supplied by Western Union Capacity also leased to American satellite corporation. Supports off shore oil rigs
SATCOM I, II (USA : RCA)	1975/76	—	24	6/4	34	—	32	8	6,000 telephony *or* 12 colour TV	Used for rented circuits and pay iTV transmission Leading role in improving communications to Alaska
COMSTAR (USA : COMSAT GENERAL) Series D1,D2,D3,D4	1976 1977 1978 1981	—	24	6/4	34	—	33	7	—	Rented by ATT and GTE for long distance circuits
SBS 1, 2 (USA : Satellite Business Systems)	1981	—	10	14/11	43	20	—	7	Transmission speeds 2.4 Kbs-6.3 Mbs Handle telephony, telex, facsimile, high speed data Teleconferencing	First system to offer all digital small-dish business services

System	Year	(kg)	No.	GHz				Life (yr)	Special payload	Remarks
TELSTAR (USA : AT&T) Series: 3A,3B,3C	1983 onwards	—	24	6/4	34	—	32	7	—	For use by AT&T
PALAPA Series (Indonesia) PALAPA A1, A2	1976/77	—	12	6/4	34	—	34	7	—	Designed to link up the 1000s of islands in the Indonesian Archipelago PALAPA 2 will in addition handle portion of telecommunications traffic within S.E. Asian countries
PALAPA 2	1983	—	24	6/4	34	—	32	8		
UNISAT (UK: United Satellites Ltd)	1986	750	2 / 6	17/12 / 14/11	27 / 36	200 / 20-30	— / —	7.5	2 Direct Broadcast TV channels / 6 Tele-communications transponders	UK domestic satellite Coverage extends to N. America and Europe To provide DBS in UK/Europe Telecommunications, small-dish business services, UK, Europe & N. America
TELECOM 1 (France)	1984	600	2 / 2 / 6	6/4 / 6/4 / 14/11	120 / 40 / 36	8.5 / 8.5 / 20	— / — / —	7	—	French domestic, also serving overseas territories Telephony, TV Small-dish business systems, also military use
TDF1 (France) TV-SAT (Germany)	1984	1,000	3	12/11	27	200	—	7	—	Direct broadcast TV to homes and community antennae

Table 3.3 Communications Satellites : National Systems (Existing, Proposed)

Name Sponsor/launch authority	Launch Date	Orbit Mass (kg)	Transponders					Life	Capacity	Comments
			No.	Frequency (GHz)	Bandwidth (MHz)	Power (Watts)	EIRP (dbw)			
OTS (Europe: ESRO/EUTELSAT)	1978	444	6	14/11				7 years		Experimental vehicle, precursor of the ECS series Various payload configurations used Variety of experiments performed
ECS-1 ECS series (Europe: EUTELSAT)	1983 onwards	550	12	14/11	80			7 years	12,000 telephony *plus* 2 TV channels	To interconnect national telecommunications networks in Europe TV distribution Later versions starting with ECS-2 will offer small-dish specialised services in conjunction with TELECOM-1
ARABSAT (22 member consortium of Arab states)	1984 onwards	680	25 1	6/4 3/2.5	33 33			7 years	8,000 telephony channels	To supply telecommunication services and community TV reception by 3 metre antennae

Table 3.4   Communications Satellites : Some Regional Systems

4 The Satellite Orbit

INTRODUCTION

It is appropriate that the discussion of the technical principles of satellite communications should start with the satellite orbit. Although we summarised the essential properties of the geostationary orbit in Chapter 2, these are not the only aspects which have to be considered. For example, there are the criteria for selecting the orbit; the advantages and disadvantages of different types of orbit; the procedures for satellite launching and positioning; and the procedures for ensuring that it maintains a stable orbit. Furthermore, the choice of orbit and the orbital position of the satellite have implications for both the satellite performance and that of its neighbours, as well as for continuity of service. In this chapter we address the following topics:

— orbital dynamics and the description of satellite orbits;

— choice of orbit, including its shape and altitude;

— satellite spacing and orbital capacity;

— satellite elevation: its implication for earth coverage and transmission link performance;

— the effects of solar and lunar eclipses and their compensation;

— launching and positioning procedures;

— satellite drift and station keeping.

ORBITAL DYNAMICS

A space satellite is in effect an artificial moon, and as such its behaviour is described by the laws of celestial mechanics. In the context of the solar system these were first enunciated by Johannes Kepler in the form of Kepler's Laws. These were eventually subsumed under Sir Isaac Newton's more comprehensive Law of Universal Gravitation and the associated Laws of Motion. These provide the analytical basis for all the calculations involved in launching and positioning space satellites.

Newton's Law is expressed in the following terms: "The force of attraction between two bodies is directly proportional to their respective masses and inversely proportional to the square of the distance between them". Or, in symbols:

$$F = \frac{M_1 M_2 . K}{d^2}$$

where K is the constant of gravitation.

This law only applies to two bodies, whereas in a planetary system, three or more bodies may be involved. Thus, in the case of an earth satellite the moon also exerts an influence sufficient to cause small deviations in its orbit. Although this could be calculated and allowed for, the simple two-body approach has been found to be adequate for the majority of space missions.

It may be helpful to examine why a satellite remains in its orbit rather than falling to earth. A popular explanation is that the centrifugal force caused by its rotation around the earth exactly balances the earth's gravitational attraction. A more accurate explanation is that the satellite's velocity, in the absence of gravity, would carry it away from the earth, but that the acceleration due to gravity counterbalances the velocity, allowing the satellite to remain in the orbital path. The closer the satellite is to the earth the stronger is the gravitational pull, and so the faster the satellite must travel to avoid falling to the earth. Low earth orbit satellites travel at about 17,500 miles/hour, and at this speed they circumnavigate the earth in about 1.5 hours. In contrast, a communications satellite in geostationary orbit travels at about 6,879 miles/hour and takes 23 hours 56 minutes 4.1 seconds to complete its circuit of the earth.

ORBITAL CHARACTERISTICS

Orbital Shape

Mathematical analysis shows that artificial satellites projected into a gravitational field in general trace out a path belonging to the family of curves known as conic sections. These comprise the ellipse, the circle, the parabola, and the hyperbola. In satellite communications the orbits of interest are either elliptical or circular. The main features of an elliptical orbit – characteristic of planetary orbits in the solar system – are illustrated in Figure 4.1.

A particular ellipse is defined by its two *foci* and the value of a quantity called the *eccentricity*. In general, an earth satellite describes an ellipse with the earth at one focus, just as the planets in the solar system describe elliptical orbits with the sun at one focus. The *apogee* is the point on the orbit where the satellite is farthest from the earth, and is travelling at its slowest speed. Correspondingly the *perigee* is the point where it is nearest the earth, and travelling at maximum speed. Using geometrical methods, it can be shown that a circle is a special case of an ellipse.

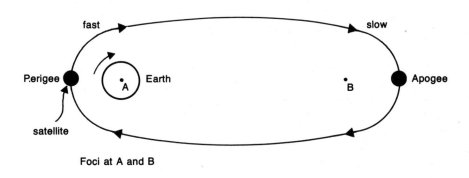

Figure 4.1   Principal Features of an Elliptical Orbit

Types of Orbit

Figure 4.2 shows three types of orbit. Both the RCA and TELSTAR satellites travelled round the earth in just a few hours, and this was their major disadvantage for telecommunications. They were within line-of-sight for only brief periods, and because they were also in motion relative

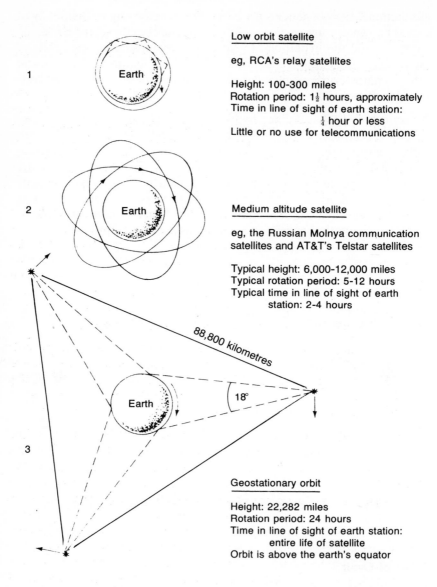

Low orbit satellite

eg, RCA's relay satellites

Height: 100-300 miles
Rotation period: 1½ hours, approximately
Time in line of sight of earth station:
¼ hour or less
Little or no use for telecommunications

Medium altitude satellite

eg, the Russian Molnya communication
satellites and AT&T's Telstar satellites

Typical height: 6,000-12,000 miles
Typical rotation period: 5-12 hours
Typical time in line of sight of earth
station: 2-4 hours

88,800 kilometres

18°

Geostationary orbit

Height: 22,282 miles
Rotation period: 24 hours
Time in line of sight of earth station:
entire life of satellite
Orbit is above the earth's equator

James Martin, COMMUNICATIONS SATELLITE SYSTEMS, © 1978, p 43
Adapted by permission of Prentice-Hall Inc, Englewood Cliffs, N J

Figure 4.2   Satellite Orbits

to the earth, the earth stations had to be capable of tracking them.

Although the USSR uses circular orbits it has continued to use medium-altitude elliptical orbits far longer than the West, and there are interesting reasons for this.

The orbits of the MOLNYA series are generally high elongated ellipses inclined at an angle of about 60° to the plane of the equator (so that they are in a roughly NE-SW direction). This offers two main advantages: first of all, the Soviet Union has no launch facilities in low latitudes; and secondly a geostationary orbit is of limited value for high-latitude regions such as Siberia. However, by having several satellites, appropriately spaced, a continuous service can be provided.

In a *geosynchronous* orbit, the period of revolution of the satellite is equal to the period of rotation of the earth about its axis. This term is sometimes employed synonymously in place of *geostationary,* but the usage is not strictly correct. A geosynchronous orbit can be circular or elliptical, and may be inclined at various angles to the plane of the equator. The geostationary orbit is unique: in addition to its geosynchronous property, the orbit is circular, it lies in the equatorial plane, and the satellite has the same direction of rotation as the earth.

The geostationary orbit is the favoured orbit for communications satellites. In circular orbit at an altitude of 22,282 miles above the equator (35,860 kilometres) and travelling in the same direction as the earth's rotation, the satellite completes its circuit of the earth in the earth's rotation time of approximately 24 hours.

Figure 4.2(3) shows that three geostationary satellites placed at the vertices of an equilateral triangle having a side equal to 88,800 kilometres can cover the entire earth with the exception of the regions close to the poles. The greater the height of the satellite, the larger the portion of earth which it covers; and, the greater the earth coverage, the greater can be the maximum permissible distance between earth stations to guarantee simultaneous line of sight, and therefore intercommunication capability. In fact, the geostationary orbit is about the optimum in this respect, and there is little benefit in placing the orbit any higher. At this height, the angle of view of the satellite is about 18° and it covers about 42.3% of the earth's surface. (This reduces to 38% if angles of elevation less than 5° are excluded, but see below.) The maximum permissible distance between earth stations is about 11,000 miles.

A variety of orbits are used for satellites serving specialised purposes other than commercial communications. One example is the LANDSAT earth resources satellite which executes a succession of polar orbits during a single rotation of the earth. Thus, on each orbit, the satellite scans and relays back to earth visual images of successive strips of the world's surface. The photographs are then interpreted and used to map the regions of the earth, to chart vegetational change, to identify geological structures, to locate mineral deposits, etc.

Advantages/Disadvantages of Geostationary Orbit

The geostationary orbit has several major advantages and one or two disadvantages, but at present the former outweigh the latter. These are the main advantages:

— because the satellite remains stationary relative to earth, the cost of sophisticated tracking equipment is avoided, thereby minimising the cost and complexity of the earth stations;

— locations within the satellite's area of coverage remain in line-of-sight contact. The break in transmission, which occurs when a non-stationary satellite disappears over the horizon, is thereby avoided;

— because of the large coverage area, a large number of earth stations can intercommunicate;

— a relatively small number of satellites can provide almost total global coverage;

— because the satellite, apart from minor drifts, experiences no motion relative to the earth station, there is almost no *Doppler shift*. A familiar example of Doppler shift is the change in the apparent pitch of the sound as a train approaches an observer and then recedes. Satellite motion relative to the earth would similarly cause a change in the apparent frequency of the radiation to and from the satellite. Satellites in elliptical orbits have different Doppler shifts with respect to different earth stations, and this increases the complexity and costs of the earth stations.

The disadvantages of the geostationary orbit are:

— latitudes greater than 81.5° North and South are not covered; and this is reduced to 77° North and South, if antennae elevations less than 5° are excluded. However, there is little else but polar ice at these latitudes;

— because of the great distance, the received signal strength, which is inversely proportional to the square of the distance, is very weak – of the order of 1 picowatt or billionth of a watt;

— the signal propagation time is also proportional to distance, and at 270 milliseconds (on average), this is sufficient for it to have a significant effect on transmission efficiency although the impact on the ultimate end user depends upon the nature of the information being transmitted and on the application;

— compared with lower altitude orbits, more powerful rocketry and on-board fuel supplies are required to achieve geostationary orbit;

— with increasing altitude the effects of earth and lunar eclipses become increasingly pronounced;

— because an antenna is not 100% efficient there is always some loss of radiated energy to space. This is referred to as the *Free Space Loss* and is the largest source of transmission loss in a satellite communications system. The Free Space Loss increases with increasing distance.

SATELLITE SPACING AND ORBITAL CAPACITY

Like any closed elliptical or circular orbit, the geostationary orbit is finite in extent, and this places a theoretical upper limit on the number of satellites that can be packed into it. In practice there are a number of considerations which influence packing density, not the least being the ingenuity of satellite designers in overcoming orbital capacity limitations.

The most important constraint is the operating frequency of the satellite. Satellites using the same frequency must not be placed too close together, otherwise their transmissions will interfere with each other. The majority of satellites currently in geostationary orbit operate in the 6/4 GHz band, for which a spacing corresponding to an angular separation of 4–5° has been found satisfactory.

How close satellites can be without interference is a complex question and depends upon a number of factors. These include:

— the transmission frequency and power of the earth stations;

— the bandwidth, frequency and power of the satellite's transponders;

— the diameter and performance characteristics of both the earth station and satellite antennae.

The interplay of these factors will be considered in later chapters. The rapid expansion of satellite communications and the requirements of new applications, particularly Direct Broadcast Television, are placing severe demands on orbital capacity. International competition for satellite slots is now a hard reality and is introducing significant political overtones into the international forums where these issues are discussed and resolved. There is accordingly mounting pressure both to reduce satellite spacing and to increase satellite transmission capacity.

The problem is being tackled in a number of ways, one of which is to use other portions of the frequency spectrum. Thus, if a satellite transmitting in the 6/4 GHz band is stationed next to one employing the 14/11 band, then the greatly reduced likelihood of mutual interference enables them to be packed more closely together.

The allocation of orbital slots and satellite frequencies is subject to international agreement, the principal forum being the International Telecommunications Union (ITU), although INTELSAT also plays an important role.

At present, orbital "slots" for satellites are provided at 3° intervals, creating a total of 120 positions round the earth's equatorial circumference. Because the demand for orbital locations is expected to exceed this number, proposals are under discussion to reduce the interval to 2°, creating an extra 60 locations.

This would reduce the distance between satellites from 2,200 km to less than 1,500 km, and it would require earth-station antennae to produce narrower microwave beams. This would ensure that a beam directed at one satellite did not spread so wide by the time it reached the orbiting position that it could cause interference with adjacent satellites.

To satisfy the new proposals it will be necessary for some satellites to be re-positioned, and also may involve modifications to earth stations. For these reasons the implementation of the changes has been phased. Table 4.1 lists the proposals.

Frequency	Stage 1 Separation	Stage 2 Separation
6/4 GHz	4°	2° (Gradually)
14/11 GHz	3°	2° May 82
30/20 GHz	To be determined	1°?

Table 4.1   International Proposals for Satellite Spacing

Table 4.2 will help to convey some idea of the projected demand for orbital slots. This is extracted from a larger table covering the complete orbit, and compiled by the ITU. The extract shows existing and proposed satellite positions for the section of the orbit extending from Longitude 19° West to Longitude 35° East. This embraces a region extending roughly from mid-Atlantic to the Western Atlantic. In examining the table, the reader should particularly note the presence of the INTELSAT satellites serving the transatlantic routes and the European regional and national satellites (located between 11.5° W and 13° E). Apart from the name or designation of the satellite, the table also lists the country of origin or ownership and the operating frequencies.

The positions for which there is an exceptionally high demand are the mid-Atlantic, the mid-Pacific and the geostationary arc serving North and South America. For each of these regions there is also a preferred position which permits the highest angles of elevation at the earth stations. For example, for Canada 100° W is favoured to facilitate communications with the far North, and for US domestic satellites the positions from 85° to 115°W will provide satisfactory elevations almost everywhere on the mainland. However, because of growing demands from S American countries and the requirements arising from the US Direct Broadcast Television plans, the capacity of the N American arc is now under very intense pressure.

ANGLE OF ELEVATION

The angle of elevation of a satellite is the angle between the earth and the

Orbital position	Space station		<1	<3	4	6	7	11	12	14	>15
19 W	F/LST	L-SAT									17
19 W *	D	TV-SAT									17/18
19 W #	F/LST	L-SAT							12	14	
18.5 W *	USA/IT	INTELSAT 4A ATL 2			4	6					
18.5 W *	USA/IT	INTELSAT 5 ATL 2			4	6		11		14	
18.5 W *	USA/IT	INTELSAT MCS ATL A		3	4	6					
18.5 W #	USA/IT	INTELSAT 5A ATL 4			4	6		11		14	
18 W	BEL	SATCOM-II					7				
18 W	BEL	SATCOM 3					7				
16 W *	URS	WSDRN						11		14	15
15 W	USA	MARISAT-ATL	1	3	4	6					
14 W	URS	LOUTCH-1						11		14	
14 W	URS/IK	STATSIONAR-4			4	6					
14 W	URS	VOLNA-2		3							
13.5 W *	URS	POTOK-1			4						
12.5 W #	F	MAROTS-B	1	3							
12 W	USA	USGCSS 3 ATL					7				
12 W	USA	USGCSS 2 ATL					7				
11.5 W	F/SYM	SYMPHONIE-2	1		4	6					
11.5 W	F/SYM	SYMPHONIE-3	1		4	6					
11 W *	URS	STATSIONAR-11			4	6					
11 W #	F	F-SAT 2		3					12	14	20/30
8 W *	F	TELECOM-1A		3	4	6	7		12	14	
6 W #	G	SKYNET	1				7				43/45
5 W *	F	TELECOM-1B		3	4	6	7		12	14	
4 W *	USA/IT	INTELSAT 4A ATL 3			4	6					
4 W *	USA/IT	INTELSAT 4A ATL 1			4	6					
4 W	USA/IT	INTELSAT 4 ATL 1			4	6					
1 W *	USA/IT	INTELSAT 4A ATL 2			4	6					
1 W	USA/IT	INTELSAT 4 ATL 4			4	6					
1 W #	USA/IT	INTELSAT 5 CONT 4			4	6		11		14	

Notes:　* } Notified
　　　# } or planned
　　All other entries are satellites in orbit

Orbital position	Space station		Frequency bands GHz								
			<1	<3	4	6	7	11	12	14	>15
1 W #	G	SKYNET 4A	1				7				43/45
0 E #	G	SKYNET-A	1				7				43/45
0 E	F/GEO	GEOS-2	1	3							
0 E	F/MET	METEOSAT	1	3							
1 E #	LUX	GDL-5				6		11	12	14	
4 E *	F	TELECOM 1C		3			7		12	14	
5 E #	S	TELEX-X		3				11	12	14	17/18
5 E	F/OTS	OTS	1					11		14	
7 E #	F	F-SAT 1		3	4	6					20/30
7 E #	F/EUT	EUTELSAT I-3						11	12	14	
10 E #	F	APEX		3	4	6					20/30
10 E	F/EUT	EUTELSAT I	1					11		14	
12 E	URS	PROGNOZ-2		3	4						
13 E	F/EUT	EUTELSAT I-2	1					11		14	
14 E #	NIG	NAT. SYSTEM			4	6					
15 E #	ISR	AMS			4	6		11		14	
16 E #	I	SICRAL 1A	1				7		12	14	20/45
17 E *	ARS	SABS						11		14	
19 E	ARS	ARABSAT I		3	4	6					
19 E #	LUX	GLD-6				6		11	12	14	
20 E #	NIG	NAT. SYSTEM			4	6					
20 E	F/SIR	SIRIO-2	1	3							
22 E #	I	SICRAL 1B	1				7		12	14	20/45
23.5 E #	D	DSF-1		3				11	12	14	20/30
26 E	ARS	ARABSAT II		3	4	6					
26 E *	IRN	ZOHREH-2						11		14	
28.5 E #	D	DFS-2		3				11	12	14	20/30
29 E	F/GEO	GEOS-2	1	3							
32 E #	F	VIDEOSAT-1		3					12	14	
34 E *	IRN	ZOHREH-1						11	12	14	
35 E *	URS	GALS-6					7				

Reproduced from ITU Booklet Number 32 by Courtesy of the International Telecommunication Union

Table 4.2 Existing and Proposed Satellite Positions on the Geostationary Arc 19°W to 35°E at 31 December 1983

line of sight from the earth station to the satellite (see Figure 4.3). The diagram also indicates that the angle of elevation varies with the location of the earth station, depending upon both its latitude and longitude. The angle of elevation has several important implications for the performance and design of satellite communication systems.

Although a satellite in geostationary orbit has about 42% of the earth's surface in view, whether or not an earth station is in line-of-sight will depend upon its angle of elevation relative to the fixed satellite position. Figure 4.4 illustrates the relationship between angle of elevation and the percentage of the hemisphere which is covered at that angle of elevation.

The angle of elevation is also related to the slant range or distance between the satellite and the earth station: the smaller the angle of elevation, the further the signal has to travel, and this has two consequences. Firstly during its journey the signal has to pass through more of the earth's atmosphere; and precipitation in the form of clouds, rain and fog

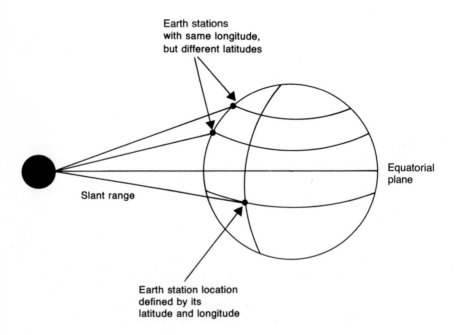

Figure 4.3   Satellite to Earth Station Elevation

can seriously impair the signal quality. This occurs in two ways: absorption of electrical energy reduces the signal strength, and scattering interactions with the water molecules introduce noise.

Besides this effect and the free space loss referred to earlier, there are other sources of signal strength loss. It is crucially important to maximise the ratio of signal strength to noise in order to maximise the information-carrying capacity of the satellite link.

The degree of severity of these effects also depends upon the transmission frequency. For the 6/4 GHz range it rises quite rapidly for elevations below 10° and for the 14/11 range, the elevation threshold increases to 30°.

The second consequence of increasing the distance travelled is that the signal transmission time or propagation delay is directly proportional to

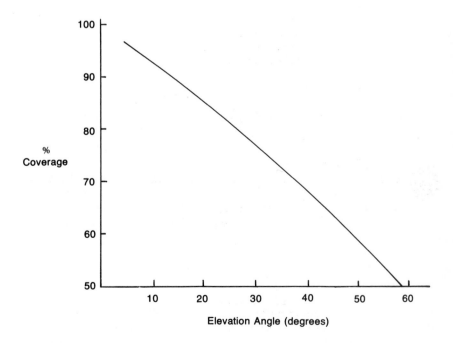

Figure 4.4   Percent Hemispheric Coverage and Elevation Angle

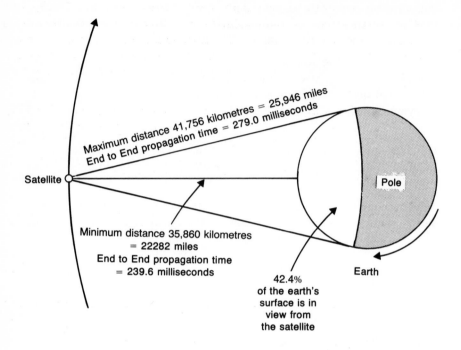

Figure 4.5   Distances and Propagation Times for a Geostationary Satellite

distance travelled. The time is calculated by dividing the distance by the speed of light – 186,000 miles/second. The minimum distance – when the satellite is directly overhead – is 22,283 miles, and the furthest distance is 25,946 miles. Figure 4.5 illustrates that in these extreme cases the earth station to earth station transmission delay can vary between 239.6 and 279 milliseconds.

ECLIPSES

Unlike terrestrial transmission circuits a satellite link has to contend with the effects of eclipses. We consider first of all a solar eclipse which occurs when the earth's shadow passes across the satellite (see Figure 4.6) and interrupts the power supply from the solar cells. This occurs on 44 nights in Spring and 44 nights in Autumn, and the maximum (or worst-case) eclipses occur at the equinoxes and last for about 65 minutes. Eclipses on other days are shorter. Therefore, if the satellite is to provide a continuous service, arrangements must be made to compensate for power interruptions. There are several solutions which can be applied, either singly or in combination.

The standard solution is for the satellite to carry storage batteries which are charged continuously from the solar cells, and these provide standby power during an eclipse. However, the battery capacity required to keep all transponders working may impose a weight penalty which may not be economically justifiable. In that case a satellite might be designed so that

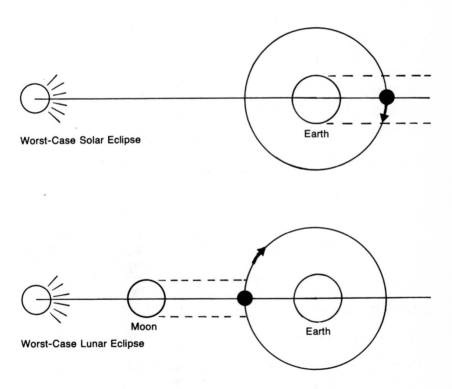

Figure 4.6   Solar and Lunar Eclipses

only a proportion of the transponders are operational for the duration of the eclipse. Much depends upon the nature of the services which are being supplied and the characteristics of the particular market. Thus, for a satellite supplying both broadcast television and data transmission services, priority might be given to the television service, if this is the primary market, and thus operating the data transmission service at reduced capacity, by temporarily taking transponders out of service.

Advantage can also be taken of the fact that when a solar eclipse does occur, the local time on earth beneath the satellite is close to midnight, and it is fortunate that this is also during a period when telephony or data traffic is low. Therefore if the satellite is positioned slightly west of the area that it services, the eclipse would then occur in the hours after midnight in the area concerned.

A more serious form of solar interference occurs when the satellite passes directly in front of the sun. Because of its very high temperature, the sun is a powerful source of noise, and this effectively blots out transmission from the satellite. The interruption lasts for about ten minutes for five consecutive days twice a year.

When this happens, the only way to maintain continuous transmission is to have a duplicate satellite in orbit, and to switch transmission channels to the standby satellite before the interruption occurs. In practice, most communications satellites do have a duplicate in orbit to ensure continuity of service in the event of failure or serious equipment malfunction.

Less commonly (about every 29 years), the moon's shadow also passes across the satellite, with effects similar to those of a solar eclipse. On occasions the moon also passes directly behind the satellite, and although the illumination from the moon does not blot out transmission to the same extent as does the sun, it does increase the noise level and significantly affects transmission performance.

LAUNCHING AND POSITIONING

The launching of space satellites has now become a commonplace affair, and satellites, designed for a variety of purposes other than communications, are despatched aloft with almost monotonous regularity. With the exception of the Space Shuttle which is the current focus of attention, few of these many launches receive much in the way of publicity. In fact, this very routineness tends to mask the extraordinary technical complexity of the operation. The task of launching and accurately positioning a satellite is one which involves the deployment of massive resources in the form of hardware and people; requires great precision; and requires the support of a global communications system.

Here we briefly review the choice of launching vehicle, launching arrangements, and the launching procedure itself.

Launch Vehicles

There is an important distinction to be made between satellite launching vehicles which can only be used once, and those which can be returned intact to earth and used again for launching satellites on subsequent space missions.

Almost all the satellites of various types currently in orbit have been launched by non-reusable multistage rockets, of which the US ATLAS CENTAUR, THOR DELTA and the European-developed ARIANE rocket are typical examples. At the present time, the sole example of a reusable vehicle is the US Space Shuttle, or Satellite Transport Service (STS) as it is sometimes called.

Such is the demand for launch capacity for satellites of all types, that both classes of launch facility are currently under great pressure. Figure 4.7 illustrates a selection of launch vehicles to the same scale.

Reusable vehicles which can be returned intact to earth offer two significant advantages. They enable the capital cost of the vehicle to be spread over a number of satellite launches and space missions; and they make it possible for a crew to be carried aloft to perform tasks and conduct experiments which would not otherwise be practicable.

The Space Shuttle

The Space Shuttle was conceived and developed by NASA partly to reduce the soaring costs of the space programme but also to further extend the scope of space exploration. The general appearance of the Space Shuttle is familiar enough but its sheer scale is not easy to fully comprehend unless one has seen the device on its launching pad. Fully assembled, with its main fuel tank in place, the vehicle's height approximates to that of the tower of Big Ben. The cargo bay of the Shuttle, about 60 feet long and 15 feet in diameter, can accommodate four satellites, together with a range of experimental equipment. It can also be fitted out as a complete space laboratory. The wide role of the Shuttle in the space exploration programme is outside the scope of this book.

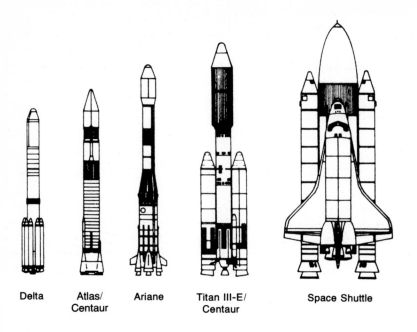

| Delta | Atlas/
Centaur | Ariane | Titan III-E/
Centaur | Space Shuttle |

Figure 4.7   A Selection of Launch Vehicles

A long-term goal for the 1990s is to use the Shuttle to build a space platform by assembling separately launched components. Such platforms will overcome the size, power and capacity constraints of separately launched satellites, and are expected to vastly extend the capabilities of satellite communications. They will, for example, enable large increases in power; make it practicable to mount very large antennae in space; and even to place very powerful computers on board.

A number of communications satellites have so far been launched using the Shuttle, and many more are planned. Table 4.3 shows a provisional utilisation schedule for the Shuttle, covering various types of missions including commercial communications satellites.

Apart from the significant reduction in launch costs, the Shuttle will also be used to retrieve satellites in order to carry out repairs, the aim being to extend a satellite's life-span. This would be effected by means of

	83	84	85	86	87	88	89	90	91	92	93	94	Total
NASA	3	4	3	6	5	7	6	7	13	12	10	8	**84**
Other Gov't (Civil)						2		1	2	1	1	2	**9**
US Commercial		3	3	3	4	3	4	5	6	6	5	6	**48**
Foreign	2	1	2	1	3	1	3	4	5	4	4	5	**35**
US Dept of Defence		1	3	5	9	10	14	15	11	14	17	16	**115**
Other		1	2	1		1	1	2	3	3	3	3	**20**
Total	**5**	**10**	**13**	**16**	**21**	**24**	**28**	**34**	**40**	**40**	**40**	**40**	**311**

Courtesy of Satellite Business Systems

Table 4.3   Space Shuttle Launch Reservations (1983)

an extending arm which would grapple for the satellite and hold it stationary. The necessary repairs could then be performed; either in space on-board the Shuttle – or the satellite could be brought back to earth. This will make it possible to extend a satellite's in orbit life-span beyond the present 7 to 10 years, and can be expected to have a major impact on communications satellite operating costs and tariffs. The feasibility of this type of operation was first demonstrated in April 1984 when a malfunctioning solar observation satellite was successfully repaired and repositioned in low earth orbit.

Expendable Launch Vehicles (ELVs)

In contrast to the Shuttle there is a wider choice of Expendable Launch Vehicles. Not only are there several basically different rockets, but there is also scope for using the same secondary stage rocket with different primary stages, as indicated by the ATLAS/CENTAUR, TITAN/CENTAUR combinations in Figure 4.7.

Since this book is concerned primarily with the European context it is appropriate that it should focus on the ARIANE launch vehicle as a typical example of expendable launch vehicles.

Until recently, countries and organisations wishing to launch satellites had to rely mainly on American launch facilities at Cape Canaveral, adjacent to the Kennedy Space Centre. However, about a decade ago the European Space Agency (ESA) embarked on the ARIANE project, the aim of which was to develop an independent European launching capability. The ARIANE rocket was designed and built by the Arianspace Company under the auspices of ESA with France having a 60% investment in the venture. Following a protracted development period and a number of teething problems – two out of six launches had failed – ARIANE succeeded in placing two commercial communications satellites into orbit in June 1983. One of these was the first European communications satellite (ECS–1), the other being AMSAT – an amateur radio satellite. This now firmly establishes ARIANE as a viable European alternative to American launch vehicles.

ARIANE is a three-stage rocket which, when equipped with the SYLDA double launch system, is capable of lifting two satellites into orbit simultaneously, compared with the Shuttle's maximum of four. When fully assembled, in the launch position and with the satellite module sitting on top, ARIANE stands about 133 feet high.

ARIANE is launched from Kouru in French Guiana. Using this location, the launch can be made due East; this takes advantage of the earth's rotation, giving a launch mass advantage of about 14%. Moreover, because it is so close to the equator, the orbital corrections following injection into the geostationary orbit are fewer and simpler.

Despite its relatively slow development and earlier technical problems, a number of organisations have already "double booked" launch capacity on ARIANE and either the US Shuttle or other expendable vehicles.

Table 4.4 lists a provisional launch programme up to 1987, and comprises both firm orders and options yet to be taken up. Amongst the satellites included in the list are several for INTELSAT; the European ECS series; and UNISAT, the UK domestic satellite.

There are two reasons for this interest in ARIANE and why organisa-

1985	January	V13 (AR-3)	Telecom-1B or SBTS-1
			G-Star 1B
	March	V14 (AR-3)	SBTS-1 or Telecom-1B
			Spacenet-3
	May	V15 (AR-1 or -2)	SPOT-1/Viking or Intelsat-V
	July	V16 (AR-1)	Giotto
	August	V17 (AR-3)	SBTS-2
			ECS-3
	September	V18 (AR-2)	TV-Sat
	October	V19 (AR-2 or -1)	Intelsat-V or SPOT-1/Viking
	November	V20 (AR-2)	TDF-1
1986	January	V21 (AR-2)	Intelsat-VA F15
	March	V22 (AR-4)	Ariane 4 - 01 (demonstration)
	May	V23 (AR-2)	Intelsat-VA F13
	June	V24 (AR-3)	Flight Opportunity
	August	V25 (AR-4)	Unisat-1 (R)
			Flight Opportunity
	November	V26 (AR-3)	STC (R)
			Flight Opportunity
	December	V27 (AR-4)	Intelsat-VI (R)
1987	February	V28 (AR-4)	Tele-X (C)
			Unisat-2 (R)
	March	V29 (AR-3)	DBSC-1 (R)
	April	V30 (AR-4)	Intelsat-VI (R)
	May	V31 (AR-3)	TDF-2 (R)
	June	V32 (AR-3 or -4)	DFS-1 (R) or Anik (R)
			Operational Meteosat-1 (C)
	July	V33 (AR-3)	Olympus (C)
	August	V34 (AR-4)	Intelsat-VI (R)
	September	V35 (AR-3)	DBSC-2 (R)
	October	V36 (AR-4)	Italsat (R)
			Rainbow (R)
	December	V37 (AR-2)	SPOT-2 (C)

Up to V23 incl.: all firm contracts   After V23: (C) = Contract   (R) = Reservation

Reproduced from the ESA Bulletin by
Courtesy of the European Space Agency

Table 4.4   ARIANE Launch Manifest 1985-1987

tions have thought fit to take out insurance policies by double booking. One is the existing and expanding worldwide demand for satellite launch facilities. The other is that for satellites having certain specialised research objectives, timing may be quite critical, and only a limited time "window" is available for the actual launch.

Now that the future of ARIANE is assured, its longer-term role and capabilities are being re-evaluated. A more powerful version is already planned, and a programme of progressive enhancement has been determined in order to ensure that the vehicle will continue to meet worldwide demand and maintain a competitive position with respect to the US.

The Choice of Satellite Transport Service

There are evident advantages to satellite operators and constructors in having not only a choice of launching facility, but also the flexibility of being able to make the final decision at a fairly late stage. In order to provide this flexibility it has been the common practice to build satellites so that they can be launched by the Shuttle or mated to one or other of the Expendable Launch Vehicles.

Although the different launch facilities are, in principle, interchangeable, there are however major differences in their capabilities and costs. For instance, rockets tend to favour longer, thinner satellites, whilst the Shuttle can more easily accommodate shorter, stubbier ones. The Expendable Launch Vehicles and different rocket combinations vary in respect of the satellite launch weights they can accommodate, and the maximum height that they can project a given weight. For example, the Shuttle, with its take-off thrust of 2,000 tons, can lift a load of 30 tons into low earth orbit. This compares with the current version of ARIANE which can launch a 1,700 kg payload into transfer orbit or 4,800 kg into circular low earth orbit. ARIANE 5 scheduled for 1990 will be able to inject loads of up to 10.5 tons into low orbit, or up to 5.5 tons into transfer orbit.

However, an important difference between ELVs and the Shuttle is that the latter is unable to inject a satellite to geostationary orbit height entirely unaided and an additional propulsion unit or PAM (Propulsion Assist Module) – also referred to as the *perigee motor* – attached to the satellite is required to lift it from the Shuttle's low earth or *parking orbit* to the *transfer orbit*. In contrast, an expendable rocket such as ARIANE can lift a satellite directly to transfer orbit height.

In the USA, launch services have become increasingly reliant on the Shuttle since it was introduced, and the trend has been accelerated by US government policies directing NASA and the military to progressively discontinue funding almost all ELV operations.

As regards the comparative costs of using the Shuttle and ARIANE it is virtually impossible to obtain accurate information and the various figures which are quoted are generally inconsistent. NASA's *price/kilogram* of satellite in 1983 using the Shuttle has been commonly quoted as being little more than half that of ARIANE. However, for a number of reasons there is a strong belief that commercial realities are likely to significantly narrow the gap over the next few years.

There is now intense competition in the satellite launching industry – particularly between the Shuttle and ARIANE – for business which is forecast to be worth many billions of pounds over the next twenty years.

Launch Procedures

Irrespective of whether a satellite is launched by an expendable or a re-usable vehicle, it must first be placed in a transfer orbit; this is elliptical in shape with its perigee at an altitude between 200 and 300 km, and its apogee on the geostationary orbit. The geometry is illustrated in Figure 4.8.

In the case of a Shuttle launch, as noted earlier, the Shuttle is not able to place the satellite in an elliptical transfer orbit unaided. Instead it is positioned in a circular parking orbit (see Figure 4.9).

In order to take it into the transfer orbit, an additional manoeuvre must take place. This is achieved using thrust generated by the *perigee motor* attached to the satellite. This motor is not required for an expendable rocket launch, since the latter injects the satellite directly into transfer orbit.

The broad sequence of events is described below.

Launch

Shuttle. The satellite with its attached perigee motor is ejected from the Shuttle, the ejection force providing sufficient velocity to carry the satellite away from the Shuttle. At the same time it imparts a small rotation to provide initial stability. The satellite is now in parking orbit which it may

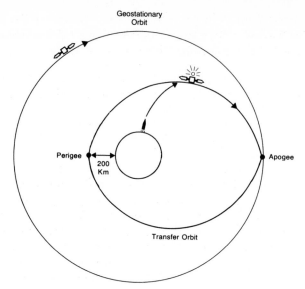

Figure 4.8   Launch Geometry – Expendable Vehicle (ARIANE)

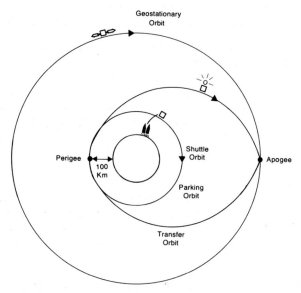

Figure 4.9   Launch Geometry – Space Shuttle

traverse several times before the PAM (or perigee motor) is fired by signals from an onboard timing mechanism to place the satellite in geostationary orbit. The perigee motor is then jettisoned.

Ariane. Ariane carries the satellite directly into transfer orbit, and the final stage of the Ariane rocket provides initial spin before the satellite separates.

Measurement and Monitoring Phase

At this point, the satellite, however launched, is in Transfer Orbit, and has some initial stability, although none of its main subsystems will be operational. In particular, the antennae are unlikely to be facing in the right direction, nor will the solar arrays be deployed. However, by drawing upon battery power and using a special omni-directional antenna (one which transmits and receives in all directions) it becomes possible to establish a communications link with tracking stations on earth. The link transmits a radio beacon, which is used initially by the tracking stations to locate the satellite in space. Once the satellite is located, the link is then used to monitor the satellite's behaviour and to issue control commands for executing further manoeuvres and subsequently to activate the various subsystems in order to achieve full operational capability.

A number of worldwide tracking networks now exist; some have been constructed for civilian purposes, others for military use. As an example the tracking network employed by NASA for the Apollo manned space programme comprised eleven land earth stations, five ships, eight specially instrumented aeroplanes and the associated cable and radio links to tie it all together and connect it to the Apollo mission control centre. Some of the communication links may themselves be provided by operational communication satellites.

Until the satellite's beacon transmission is received, the satellite is out of contact with earth, and during this period, the atmosphere in a mission control centre is one of quiet expectancy until a voice rings out with the message "we have acquisition!" (or some similar phrase). This signifies that some tracking station has picked up the beacon transmission and located the satellite.

During this phase the satellite might travel round the transfer orbit a number of times. Simultaneously, on the ground, extensive calculations are performed to establish, among other things, the appropriate time for injection into geostationary orbit.

Apogee Motor Firing

The geostationary orbit is measured as accurately as possible, and the satellite's position and orientation adjusted ready for the next step. At the precise moment when the satellite is at the furthest point of its ellipse, and travelling approximately at right angles to the earth's radius, the apogee motor is fired to put it into geostationary orbit.

Adjustment Phase

The satellite's velocity is then adjusted to synchronise with the earth's rotation, and also its orientation so that the antennae point in the correct direction.

The above description is very much an over-simplification of the operation. For example: the apogee motor may need to be fired on several occasions to achieve correct positioning and speed; also, because there are no launching sites exactly on the equator, the satellite must also be manoeuvred so that it is in the equatorial plane; and, apart from activating all the various subsystems, extensive checks are performed to make sure they are functioning correctly.

The launch of a satellite requires the support of a worldwide chain of earth stations, so that once it is capable of receiving and transmitting signals, its position and behaviour can be monitored and control commands issued from earth.

Because a satellite has a finite lifespan, and can also develop equipment malfunctions, satellites which are intended to provide a commercial service are usually duplicated in orbit, so that continuity of service can be preserved. It is also now a common practice to provide a second level of standby by having another spare on earth.

SATELLITE DRIFT AND STATION KEEPING

Satellites need to have their orbits adjusted from time to time because although the satellite may initially have been placed in the correct orbit, natural forces induce a progressive drift.

The drift from a geostationary position is caused by a number of factors: minor gravitational perturbations due to the sun and the moon; the fact that the earth is not a perfect sphere; and the pressure of solar radiation. The effect of the sun and the moon acting in a North/South direction is to cause the orbit to move away from the equatorial plane. In addition, the moon induces a daily "tidal" oscillation in the satellite's position. Also, because the earth is not perfectly spherical the gravitational field is not spherical either. This produces a gravity gradient in the circular geostationary orbit, the two "hollows" being at longitudes 79° E and 101° W. A satellite placed in these positions would be more stable than anywhere else, but most are in other positions. A velocity increment of about 7 feet/second/year is reckoned to be sufficient to compensate for this effect.

The effect of solar radiation pressure varies with the size of the satellite and the area of its solar panels, but the drift is small compared with that caused by gravitational influences.

The necessary orbital adjustments are carried out either by releasing jets of gas, or by firing small rockets positioned around the body of the satellite. The frequency with which the adjustments are made depends upon how accurately the satellite's position has to be maintained. Until recently, the tolerance was such that it only needed to be done periodically. However, the increased orbital crowding, the trend towards high-speed digital transmission and the use of Time Division Multiple Access (TDMA) techniques is imposing more stringent station keeping accuracy. Adjacent satellites must be prevented from drifting too close together to minimise radio interference, whilst digital transmission and TDMA techniques rely on extremely accurate timing of the signals between the satellite and each individual earth station.

A change in the satellite's distance or range results in a change in the signal propagation time, and, even though this may only be of the order of a few nanoseconds, it may be sufficient to cause serious synchronisation problems. Thus, the "tidal" movement referred to earlier can cause the satellite to move towards or away from earth at speeds up to 2 feet/second, so that in the space of one second the propagation time can change by up to two nanoseconds. Furthermore, although the choice of the geostationary orbit minimises the Doppler effect, uncontrolled meanderings by the satellite may increase its significance.

5 Satellite Construction

INTRODUCTION

In Chapter 2 we briefly reviewed the principal functional components of a communications satellite. In the present chapter we discuss these and other aspects of satellite design and construction in a little more detail. In particular, we concentrate on:

— antennae;

— transponders;

— power supply;

— stabilisation.

Figures 5.1 and 5.2 illustrate the overall construction and the main components of the INTELSAT IV and ECS–1 satellites. These represent the two broad categories into which satellites can be subdivided; those which have a roughly cubical (or polygonal) shape and are *axially stabilised,* and those which are cylindrical and are *spin stabilised.*

Satellite design and construction present several unique problems. First of all, the very inaccessibility of a spacecraft for repair purposes imposes stringent reliability requirements. Secondly, like any other piece of equipment it also has a limited life-span and for obvious economic reasons it is desirable that this is maximised. Finally, it has to operate effectively and efficiently in the distinctly hostile environment of space, and this poses quite novel engineering problems which dictate equally novel solutions.

The design and construction of a communications satellite is a major undertaking and involves a wide range of scientific and engineering

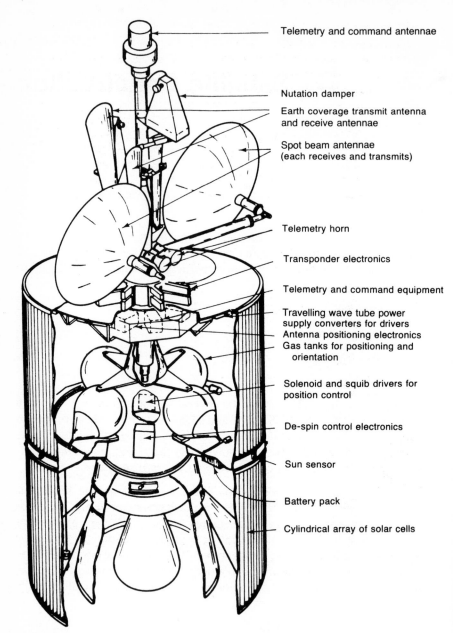

Telemetry and command antennae

Nutation damper

Earth coverage transmit antenna
and receive antennae

Spot beam antennae
(each receives and transmits)

Telemetry horn

Transponder electronics

Telemetry and command equipment

Travelling wave tube power
supply converters for drivers
Antenna positioning electronics
Gas tanks for positioning and
    orientation

Solenoid and squib drivers for
position control

De-spin control electronics

Sun sensor

Battery pack

Cylindrical array of solar cells

Figure 5.1   INTELSAT IV

disciplines. Indeed, few if any industrial organisations have the total capability to tackle a project of this scale, and it is customary to award the contract to a prime contractor who in turn appoints a number of principal contractors and subcontractors to handle the different components and subsystems. An approach which is being increasingly favoured, particularly in Europe, is for a number of companies with the appropriate mix of skills to establish consortia groupings having the total capability. As a result, satellite projects are often international, involving contributions from organisations in various countries. Thus, members of the United Kingdom and European aerospace industry, apart from acting as prime contractors, designing and building complete satellite systems, also

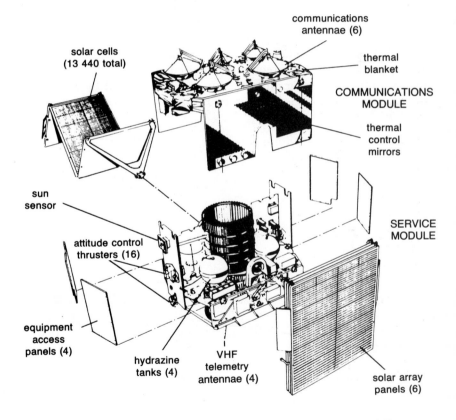

Courtesy of British Aerospace

Figure 5.2 Exploded View of the ECS-1 Satellite

develop and produce components for the US Space Shuttle, the INTEL-SAT and INMARSAT satellites, and a wide range of other spacecraft.

THE ANTENNAE SUBSYSTEM

General Principles

The earth subtends an angle of approximately 18° at a satellite in geostationary orbit (see Figure 5.3). If we now imagine a simple satellite-mounted aerial, radiating energy equally in all directions, a high proportion of the radiated energy is lost to space. Such an aerial is referred to as *omnidirectional,* and the radiation as *isotropic,* ie it radiates equally in all directions. The power that is lost to empty space is called the *free space loss*.

It is evident that, in order to increase the proportion of transmitter power actually transmitted to earth, some means of concentrating the radiation in a preferred direction is required. At microwave radio frequencies this increase in directionality can be achieved by employing antennae with dish-shaped reflectors. Figure 5.4 shows two dish anten-

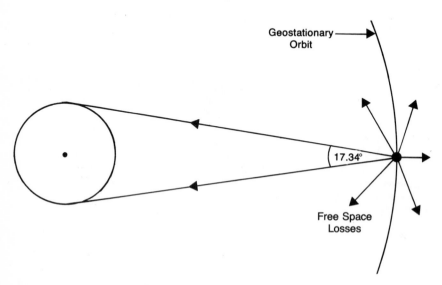

Figure 5.3   Free Space Loss

nae, one with a single feed and the other with multiple feeds. Until fairly recently, almost all antennae have been designed with parabolic cross-sections – as in a car headlamp reflector, the feed to the reflector being located at the focus of the parabola.

A parabolic reflector has the following properties: received rays entering the dish parallel to the feed axis are reflected and converge at the focus; and transmitted rays emitted from the focus are reflected and emerge parallel to the feed axis. However, in order to meet increasingly

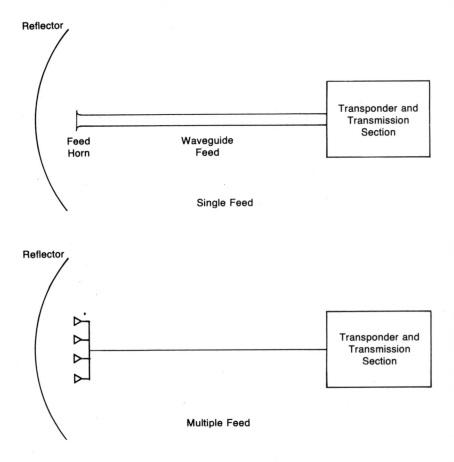

Figure 5.4   Multiple Feed Dish Antennae

exacting performance requirements, designers are now employing non-parabolic cross-sections.

Although antennae dishes frequently have a circular cross-section or *aperture* viewed perpendicular to their axis they need not be so; and elliptical and other shapes are also used.

The greatly amplified electrical signal from the transponder and its associated electronic circuitry is fed to the reflecting dish by a special kind of conductor in the form of a precision hollow tube or *waveguide*. A waveguide is used because conventional solid conductors do not conduct electrical energy efficiently at very high frequencies. The end of the waveguide adjacent to the reflector is commonly referred to as the *feed horn*.

A dish antenna can act as both a transmitter and receiver, the feed horn illuminating the reflector when transmitting, and collecting radiation when receiving. Satellites may operate in this dual mode or may function solely as transmitters or receivers; and both types may be present on the same satellite.

Antennae Performance: Gain and EIRP

It is convenient at this point to introduce two quantities or figures of merit which are used for describing the performance of an antenna. These quantities, together with several other figures of merit which will be introduced when appropriate, are of fundamental importance in the design of satellite communication systems. These quantities are: Gain (G) and Equivalent Isotropic Radiated Power (EIRP).

Gain

Gain is a measure of the directionality of an antenna, or of its power to concentrate energy in a preferred direction. The gain of an antenna is directly proportional to the aperture area of the antenna (or the square of the diameter, in the case of a circular antenna) and the square of the transmission frequency. The gain, in common with a number of other performance figures of merit, is generally quoted in *decibels*. (This unit of measure which has wide application in communications engineering is defined and explained in Appendix A.)

Gain can be used to describe both transmitting and receiving antennae, since the same antenna transmits and receives with the same gain. Since the value of the antenna beam angle, when squared, is roughly proportional to the aperture, antennae can also be compared in terms of their respective beam angles. For example a spot beam antenna covering a 4.5° angle has a gain of $(17.34/4.5)^2$ or 14.85 times that of a 17.34° earth coverage antenna.

An important result of the dependence of gain on both aperture and transmission frequency should be noted: by moving to higher frequencies the same gain can be achieved using a much smaller antenna. This is the major reason why the move from the 6/4 GHz frequency range to 14/11 GHz and upwards enables much smaller dishes to be employed at the earth stations.

Equivalent Isotropic Radiated Power

Gain, in isolation, merely describes the amplifying power of the antenna; it does not take into consideration the actual transmitter power delivered to it. This is achieved by multiplying the transmitter power by the gain to give the EIRP. Thus, a 10,000 watt transmitter and an antenna with a gain of 5,000 would result in an EIRP of 50 million watts. EIRP can also be regarded as the transmitter power which would be required in conjunction with an *isotropic* antenna to give the same result as the transmitter and antenna configuration under consideration.

EIRP is used to describe and measure transmission power both at the satellite and at the earth station, and it is here that a possible source of confusion can arise. It should be stressed that, *although EIRP is defined and measured at the transmitting antenna, its primary significance is at the receiving end of the transmission link*. EIRP measures the transmitted power at the satellite (or the earth station); received power at the other end of the link can be computed under a variety of conditions by taking the EIRP and applying to it the various gains and losses that can occur along the transmission path. Like gain, EIRP is of fundamental importance to the operators and designers of satellite communications systems. For example, both the magnitude and distribution of the satellite EIRP over its area of coverage are major determinants of earth station design.

Types of Antennae

The antennae on the first geostationary satellites were not very directional, and much of the signal power transmitted was wasted; Early Bird, for example, had only one omnidirectional antenna. Since then, not only have there been vast improvements in antennae gain and directionality, but different types of antennae have been introduced to give enhanced capabilities and to meet various service requirements. For both practical and economic reasons it is desirable that communications satellites should be able to support multiple services; for example: broadcast television, telephony, data transmission and, increasingly, small-dish business services. There may also be a service requirement to provide diffential coverage for different service regions in terms of geographical area and in terms of the power transmitted into the area. And finally it is obviously desirable to have the flexibility to be able to reallocate a satellite's capacity over different routes in order to adapt to changing traffic patterns and demands.

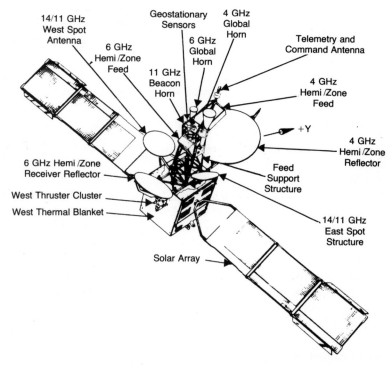

James Martin, COMMUNICATIONS SATELLITE SYSTEMS, © 1978, p 87
Adapted by permission of Prentice-Hall Inc, Englewood Cliffs, N J

Figure 5.5   INTELSAT V Showing the Various Antennae

Nowadays most communications satellites are equipped with several antennae; a large high-capacity satellite such as INTELSAT V (Figure 5.5) bristles with antennae of various sizes, shapes and configurations. The following are the main categories of antennae:

— omnidirectional;

— global or earth coverage;

— hemi/zone;

— spot beam.

Omnidirectional

Although an omnidirectional antenna provides earth coverage, it differs from an earth coverage antenna proper, because it lacks directionality. However, all satellites carry an omnidirectional antenna for use following injection into the parking orbit, before it is finally positioned. Until the satellite's directional antennae are fully deployed and oriented in the correct direction, a non-directional antenna is the only practicable means of establishing a communications channel for the Tracking, Telemetry and Command (TTC) system. This system is described below (see p 124).

Global or Earth Coverage

INTELSAT III was the first satellite to have a directional earth coverage antenna, aiming the beam down the 17.34° angle subtended by the earth, and Figure 5.6 shows the three types of antennae considered so far employed on its successor, INTELSAT IV. Earth coverage antennae are commonly referred to as horn antennae, from their shape, and are diminutive in size compared with the other types of antennae.

Hemi/Zone

In terms of their respective areas of coverage, the distinction between hemi and zone antennae is mainly a matter of degree, the former providing greater coverage than the latter. The "hemi" terminology originated in the INTELSAT context, the special significance of the hemispherical beams being that they service specified geographical hemispheres. Figure 5.7 illustrates the approximate areas of coverage of the antennae on board the INTELSAT V satellites serving the Atlantic, Pacific and Indian Ocean regions.

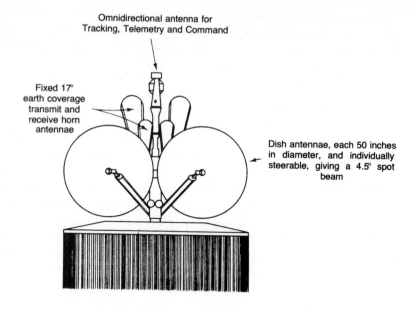

Omnidirectional antenna for
Tracking, Telemetry and Command

Fixed 17° earth coverage transmit and receive horn antennae

Dish antennae, each 50 inches in diameter, and individually steerable, giving a 4.5° spot beam

Figure 5.6   INTELSAT IV Antennae

It is significant that the contours which approximately define the areas of coverage (see Figure 5.7) are irregular, signifying that the beams have been deliberately shaped so that the area of coverage approximates to that of a particular geographical area. The technique is commonly used for beams with zonal coverage, and distinguishes them from the simpler earth coverage beams discussed above.

Spot Beams

As the INTELSAT traffic increased, it became evident that the bulk of the communications traffic was between North America and Europe. As a result INTELSAT IV, in addition to the earth coverage antenna of INTELSAT III, was equipped with spot beam antennae. Spot beam antennae employ larger reflectors than the other types of antennae, and this enables the power to be concentrated into a much narrower beam, illuminating a correspondingly smaller area on earth. Because the total transmitted power is concentrated over a much smaller area, the received signal strength at an individual location will also be higher. This produces two major benefits: it enables power and therefore transmission capacity to be concentrated on those areas with greatest demand; and the higher received signal strength also enables less sensitive and less costly earth stations to be used.

Compared with the 17.34° beam angle of INTELSAT III's earth coverage beam, the later INTELSAT V spot beams have an angle of 4.5 degrees, with the approximate coverage areas shown in Figure 5.7. Even narrower spot beams have been employed on some communications satellites, a good example being the NASA ATS–6 (Advanced Technology Satellite). This has a beam width of 1° and the earth coverage is only about 500 miles across. To achieve this requires an on-board antenna which is about 10 metres in diameter. Since it would be impracticable to launch a satellite with reflectors of this size having the normal rigid construction, a collapsible structure – looking very much like a folding umbrella – is employed. This is stored initially in its collapsed form, and

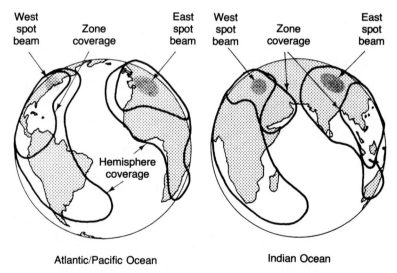

James Martin, COMMUNICATIONS SATELLITE SYSTEMS, © 1978, p 30
Adapted by permission of Prentice-Hall Inc, Englewood Cliffs, N J

Figure 5.7   Beam Coverage Areas for the INTELSAT V Satellites serving the Atlantic, Pacific and Indian Ocean Regions

then fully extended when the satellite is in orbit. With reflectors of this size it also becomes possible to design the feed arrangements so that instead of producing just a single spot beam, the antenna generates a multiplicity of distinct high-powered beams.

An important motivation for developing yet more powerful transmission beams and antennae configurations is to reduce earth station size and cost; the principal commercial pressures at the present time stemming from the requirements of direct broadcast television, small-dish business services, and third world countries. Spot beams also produce other benefits: they enable the satellite's transmission frequencies to be *re-used,* thus increasing its transmission capacity; and by interconnecting different spot beams, possibly in conjunction with spot beams that can be steered or redirected, they greatly extend service flexibility.

Steerable Antennae

The position and orientation of a satellite in the geostationary orbit can be changed – which may be necessary for various reasons. However, most communications satellites have been designed with antennae that maintain a fixed direction with respect to earth throughout the satellite's lifetime. A fairly recent innovation is to mount on-board the satellite one or more spot beam antennae whose direction relative to earth can be altered by remote command.

The zone antennae of INTELSAT IV and subsequent INTELSAT satellites are steerable and the European L–SAT will carry a cluster of five antennae which are steerable as a group. The principal advantage of steerable antennae is that they enable the beam direction and area of coverage to be altered in order to meet changing demands and traffic patterns.

An example drawn from the military field is shown in Figure 5.8, which represents the US Defence Satellite Communications System. Four satellites in geostationary orbit provide the long-haul links in the American worldwide military communications network, which links the National Commanders in Washington with all the worldwide subordinate elements of the command structure. A continuous global link is provided by a sequence of overlapping spot beams together with an earth coverage beam. This arrangement, together with the capability of being able to re-direct the beams, provides great operational flexibility, enabling new communications paths to be established at very short notice.

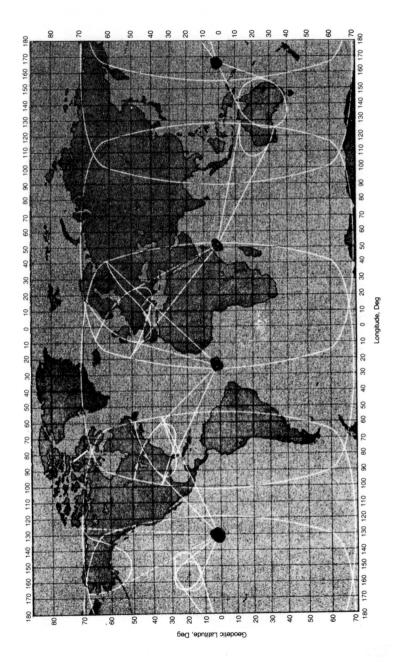

Figure 5.8 The American Defence Satellite Communications System

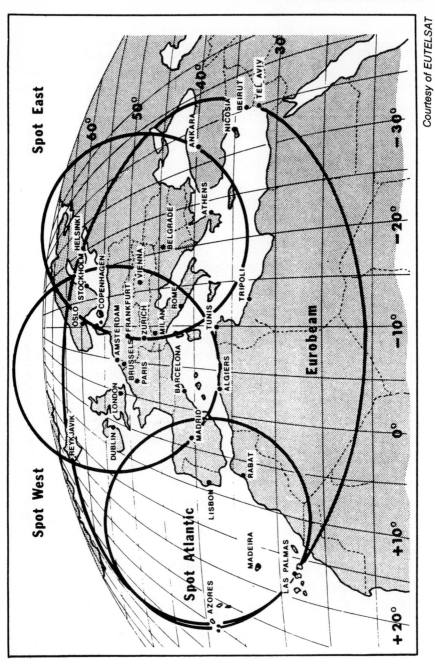

Figure 5.9 ECS-1 Satellite Antennae Coverage

Area of Coverage: Footprints and Beam Shaping

The area of coverage of an antenna beam is commonly referred to as the *footprint,* and its shape is determined by: the antenna design; the resulting beam angle and shape; and the angle of elevation of the beam with respect to earth. On the other hand, the overall specification and design of the antennae subsystem is largely determined by considerations such as the geographical characteristics of the region to be covered, the range of services to be offered together with the estimated traffic volumes and overall system economics.

Figure 5.9 shows the antennae for the European ECS–1 satellite beams. ECS–1 is designed to handle television signals and telephony type traffic and is equipped with five antennae. Of these, one is used solely for reception of *all* signals emanating from an area corresponding to the Eurobeam, and the other four are used purely for transmission. One of the transmission antennae is dedicated to television transmission over an area corresponding to the Eurobeam, whilst the other three provide spot beams providing telephony services to their respective areas. These are designated: Spot Atlantic, Spot West and Spot East.

Figure 5.9 is an over-simplification, because the received power is not distributed uniformly over the footprint, but decreases progressively as we move from the centre of the beam outwards. Furthermore, the characteristics of the market to be served, and the types of service to be offered, are amongst a number of considerations which have a strong bearing on the required received power distribution and the minimum power received at some specified boundary. For example, close to the centre of the beam, reception may be perfectly adequate using an earth station with a 5-metre dish, whereas closer to the boundary a 7-metre dish may be needed for adequate reception. For these reasons, antennae footprints are specified and computed by service operators and satellite designers, in terms of various performance criteria or figures of merit such as gain or EIRP. These can be conveniently represented by contour "maps" (see Figure 5.10, showing the footprint of the broadcast television beam for the proposed UK UNISAT domestic satellite).

In order to make the most efficient use of the transmitted power, it may be desirable to shape the satellite beams so that their footprints correspond approximately to the shapes of the countries or regions which they serve. This is obviously important where a country has a thin irregular shape.

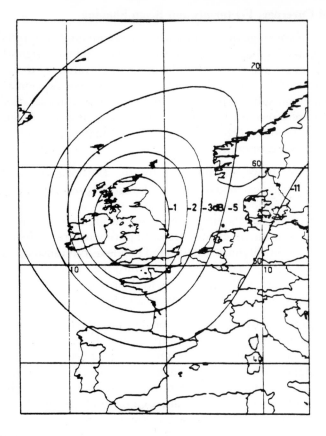

Figure 5.10   EIRP Contours for the UK UNISAT Broadcast TV Beam

Beam shaping can be achieved in several ways, the two principal methods being: to shape the antenna dish; and to direct the signal at the reflector with appropriately positioned multiple feed horns. The result of using multiple feed horns is a set of overlapping spot beams which have an envelope approximating to the desired shape, as shown in Figure 5.11. A good example is INTELSAT V which uses 88 feed horns to drive its hemi/zone reflectors to produce the desired footprint shapes.

The multiple spot beam technique offers great flexibility. For instance, it enables the beam shape to be changed by switching on or off appropri-

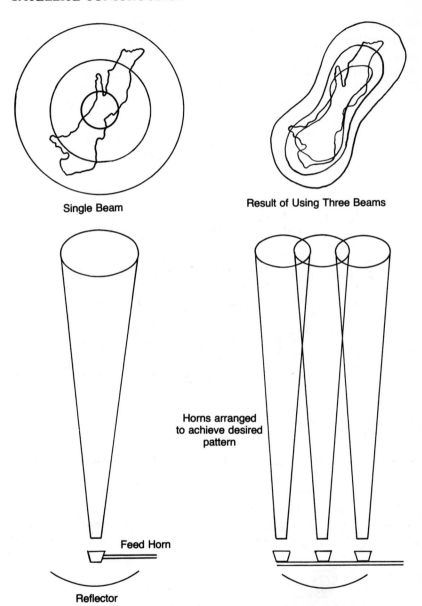

Figure 5.11 Beam Shaping Using Multiple Spot Beams

ate configurations of feed horns by remote command from earth. Detached land masses can be served by using separate feed horns with the same reflector to direct beams into those areas. These facilities are available on INTELSAT V. One final example of beam shaping is shown in Figure 5.12, which shows the predicted gain contours for the European and North American telecommunications antennae beams of the UK UNISAT satellite. The task presented to the design engineers was to develop antennae which would produce footprint contours approximating to the polygonal shapes formed by connecting together the specified cities and towns.

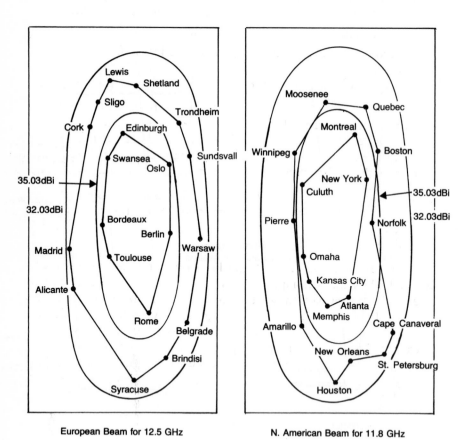

European Beam for 12.5 GHz   N. American Beam for 11.8 GHz

Figure 5.12   Gain Contours for UNISAT Telecommunications Antennae

Frequency Re-use

In the face of increasing competition for space on the geostationary orbit and of continually increasing demands on the frequency spectrum, it is essential that the frequency bands allocated to satellite communications are efficiently utilised. There are three main ways of achieving this goal: to arrange for the satellite to re-use all or part of the frequency band available to it; to employ efficient user access methods and capacity sharing schemes; and to use modulation, information encoding and compression techniques which enable more information to be packed into the radio frequency bandwidth. Here we review frequency re-use methods, leaving the other techniques for consideration in a later chapter.

The two methods in common use today are beam separation and polarisation. The beam separation technique relies on the fact that if two beams illuminate different regions without overlapping, then each beam can employ the same frequencies without mutual interference.

Polarisation on the other hand makes use of the principle that electromagnetic waves generally can be made to vibrate uniformly in one of several planes or directions of polarisation. The technique enables a single range of frequencies to be re-used by arranging to transmit (and receive) with a different polarisation to that employed for the initial use.

On-board the satellite and within the earth station, polarising circuits are used to distinguish between and separate out the different vibrations, in much the same way that polarised sun-glasses can eliminate light reflections.

Each technique, used independently, doubles the transmission capacity, and when the techniques are used in combination the effect is to quadruple it. However, these very substantial gains in effective capacity are not achieved without a cost and weight penalty, since increased power in the form of additional transponders is required in order to utilise the extra capacity. The progressive exploitation of frequency re-use techniques accompanied by increased power are the main factors responsible for the continuing increase in satellite capacity which has occurred over the years.

THE TRANSPONDER SUBSYSTEM

The transponder is essentially a *repeater* which receives a signal transmitted from earth on the up-link, amplifies the signal and retransmits it on

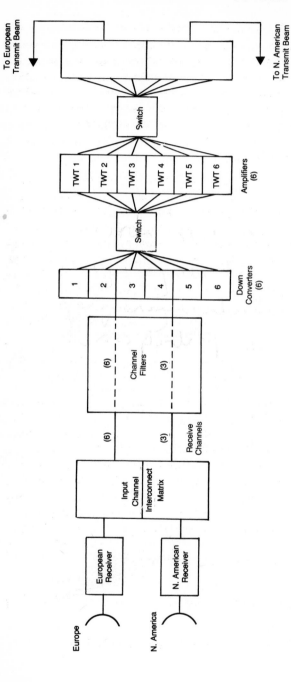

Spot Beam Frequency Allocations (Giga Hertz)

	Up	Down
Europe	14.0 – 14.25	12.50 – 12.75
N. America	14.25 – 14.50	11.75 – 12.00

Figure 5.13   Transponder Schematic

the down-link at a different frequency from the received signal. However, before proceeding any further we should note one potential source of confusion. The term *transponder* generally refers to the whole of the subsystem, including the receiver, the high-power amplifiers, and other substantial electronic circuitry. The plural *transponders,* however, refers to the amplifiers, which may share other parts of the system. Thus the terms *amplifier* and *transponder* may be used interchangeably.

A Typical Transponder Subsystem

Satellite transponder systems exhibit wide design variations, but do have certain basic features in common; Figure 5.13 shows a typical arrangement – which derives from the specification of the telecommunications package for the United Kingdom UNISAT satellite. The telecommunications services are supplied through two spot beam antennae operating in dual transmit and receive mode, one serving Europe and the other serving North America. Also, six transponders are available, each having a bandwidth of 36 MegaHertz. Apart from providing transmission links between locations served by the same spot beam, it is also planned to provide links between locations in North America and Europe served by different beams.

The up- and down-links of each beam are allocated specified frequency bands within which the corresponding earth stations must operate.

From this brief description it may be evident that, apart from its primary amplification and frequency conversion functions, the system must also perform several other tasks. First of all, because the total received bandwidth is considerably greater than that of an individual transponder, a mechanism is needed for subdividing the incoming up-link frequency bands according to some logical plan, so that the bandwidth can be spread across the transponders. Next, the signals originating on a specified receive beam must be directed to the transmit beam serving their eventual destination. Finally, the signals originating from different earth stations must be aggregated into the frequency bandwidth of the corresponding down-link, before actual transmission.

How this is achieved within the context of an overall satellite communication system is described in the next chapter. Here we note the following principles that are employed:

— the allocated up- and down-link frequency bandwidths are subdivided using the bandwidths of the transponders as a basis;

— the bandwidth subdivisions enable a set of connected paths or channels to be defined between the input antennae and the output antennae;

— although in the present generation of satellites the paths are predefined, provision is made for the beam interconnections to be changed, if required, by means of remotely controlled on-board switching units.

With these points in mind, we review the main components of the system shown in Figure 5.13.

Receiver Section

Partly for technical reasons and partly for operational reasons, independent receivers are employed for the European and North American beams, and this distinction is maintained throughout the remainder of the subsystem.

The receiver is a complex device and performs a number of functions. These include one or more stages of pre-amplification and frequency changing, and filtering operations to remove noise and generally improve the quality of the received signal.

Interconnection Matrix and Channel Filters

In this section, the receiver output bandwidths are subdivided into separate receive channels, with a separate filter chain for each channel.

It will be observed that the number of possible channels (9) exceeds the number of transponders (6). The discrepancy arises from the proposed service plan for the routes between Europe and North America (this is discussed in Chapter 6).

The Down Converters

The separate channels are next converted to their appropriate down-link frequencies depending upon the beam to which they are eventually destined. Each channel is then input, by means of a switch, to an individual transponder. (In Figure 5.13, TWT stands for Travelling Wave Tube, the particular type of High Power Amplifier device which is generally employed.)

In practice there are a number of switches present, serving various purposes. These include: altering the channel/transponder assignments; and switching in spare transponders to replace faulty ones. For our present purposes we assume that switches which change equipment and path interconnections are actuated by remote command from the ground.

Output Multiplexers

The transponder outputs are directed to either of two multiplexers serving the European and North American beams respectively. Within each multiplexer, the signals from the separate channels are then combined into a signal within the bandwidth allocated to the corresponding downlink, and the signal is then routed to the appropriate antenna feed.

High Power Amplifiers

A High Power Amplifier needs to be very reliable, to be light in weight, to possess high amplification efficiency because of the small power supply available, and to be able to operate over a wide range of frequencies. There are two devices which qualify at the present time, the Klystron oscillator and the Travelling Wave Tube (TWT). Of these, the former is sometimes used in low power earth stations, but the TWT is the most strongly favoured, particularly for the satellite end of the link.

The TWT externally resembles a tube, about 2 feet long. It carries an electronic gun which shoots a beam of electrons down the tube, the beam being focused on a collector by means of concentric magnets. The radio frequency signal is input to a wire helix concentric with the beam, generating a travelling magnetic wave in the process.

As the travelling wave moves down the tube, interaction between it and the electron beam results in a progressive increase in the strength of the signal in the helix and this occurs at an exponential (or compound interest) rate. At the end of the tube, the amplified signal is fed from the helix to the output. In exploded or cutaway views of a satellite, the TWTs can frequently be readily identified from their radial arrangement below the platform carrying the antennae (see Figure 5.1). The TWTs currently in use generate power outputs ranging from 6 watts to 200 watts. However, the greater the amplification and power output, the greater is the input power required to operate the TWT, so that the satellite's power supply is a significant limiting factor.

Transponders of various bandwidths are employed – 36, 45 and 54 MHz being typical; of these, 36 MHz has been particularly popular, a major reason being that it will handle a single broadcast colour television channel. However, transponders having wider bandwidths are being increasingly used, and transponders with different bandwidths may be present on the same satellite. Whereas the early satellites such as Early Bird had only one or two transponders, now 10, 12 and 24 are usual for a small-to-medium-capacity satellite; the number may be as high as 50 for the large INTELSAT class.

Despite the remarkable success of the Travelling Wave Tube, it does have a number of limitations. These include weight, physical size and input power requirement, but a major limitation is the non-linearity of its performance. Expressed in simple terms, the non-linearity results in inefficient usage of the available bandwidth and also introduces distortion into the signal when the TWT is driven close to its maximum power. An interesting fact about TWTs is that, although they are designed and tested to very stringent specifications and are available as standard components, they do nevertheless exhibit individual variations under operational conditions – much as the original glass thermionic valves did. Accordingly they are given an initial "burn in" period before being brought into service.

For some time there has been continuing intensive research aimed not only at improving the performance of TWTs, but also at developing a superior substitute. The potential of solid-state devices is being explored, but so far this has not resulted in devices with the necessary high power outputs.

POWER SUPPLY SYSTEM

Apart from the on-board batteries which provide standby electrical power, the sole power source is that derived from the radiant energy of the sun. The principal functions of the power supply system are to collect the solar energy; transform it into electrical power; and to generate and distribute electrical currents, with the appropriate characteristics, to the other satellite subsystems and components.

Solar Energy

At geostationary altitudes, a significant quantity of the sun's energy reaches a square metre of satellite surface perpendicular to the sun. This volume of energy is about 1,000 watts, which is almost 1.5 horsepower – a useful amount of energy if this could be captured and converted at a reasonable cost. The solar cells which carry out the conversion are made of specially doped silicon crystals about two centimetres square. These are assembled into panels or arrays and connected together to satisfy the power requirements of the satellite. To generate 100 watts, 2000 such cells may be needed, occupying about a square metre of surface.

Two Types of Solar Array

There are two forms of solar array: the solar cells can be attached to the body surface of the satellite; or alternatively assembled onto flat surfaces or solar "sails" extending from the satellite (see Figure 5.14).

Each method has its advantages and disadvantages. The satellite body surface arrangement was the first to be employed and was relatively straightforward to design for the early satellites, almost all of which were cylindrical and spin stabilised. However, cylindrical arrays are not particularly efficient because only half the cells are in direct sunlight at any instant and most of these are not perpendicular to the sun's rays. As a result, only about $1/\pi$ or approximately 30% of the incident radiation is available for conversion. This imposes a significant power limitation on satellites employing cylindrical arrays.

Flat panel solar arrays provide a more efficient configuration, but also introduce additional complications. First of all, the solar sails may each have a length of 30 feet or more and must be stored in a folded-up position during the launch stage. When the satellite reaches geostationary orbit they are then opened in a concertina fashion to their fullest extent. Also, a mechanism is required to ensure that the arrays are always pointing directly at the sun so that they receive maximum incident illumination. This is achieved by positioning sun sensors on the sails. The signals from the sensors are fed to stepping rotors which rotate the sails about their longitudinal axis, so that they follow the sun.

Flat panels which constantly face the sun also suffer far more from overheating than cylindrical arrays which are continually spinning with the satellite body. It is therefore necessary to control the cell temperatures in order to limit mechanical distortion of the sails which could be caused by differential thermal expansion.

Both types of solar array are currently in use. However, with the trend

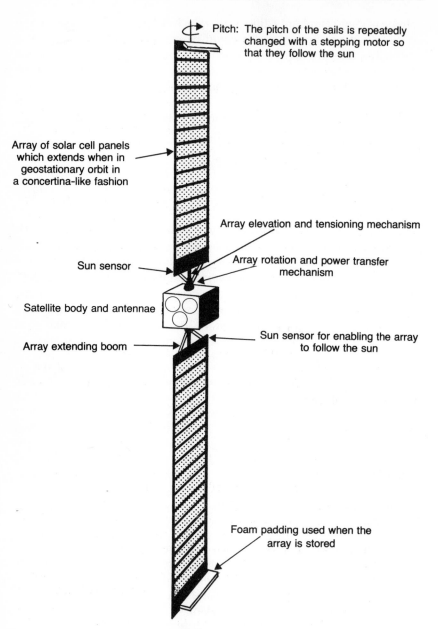

Pitch: The pitch of the sails is repeatedly changed with a stepping motor so that they follow the sun

Array of solar cell panels which extends when in geostationary orbit in a concertina-like fashion

Array elevation and tensioning mechanism

Sun sensor

Array rotation and power transfer mechanism

Satellite body and antennae

Sun sensor for enabling the array to follow the sun

Array extending boom

Foam padding used when the array is stored

Figure 5.14   Satellite with Solar "Sails"

towards more powerful satellites, the flat panel solution with extended sails attached to bodies of varying shapes and employing three axis stabilisation is increasingly favoured.

STABILISATION AND ATTITUDE CONTROL

The functions of the stabilisation and attitude control system are two-fold: to ensure that the satellite body maintains a stable orientation to the earth; and to maintain the pointing accuracy of the antennae. In the absence of such a system, an orbiting satellite would tend to "tumble" about in an uncontrolled fashion.

A stabilisation system comprises mechanical and electrical assemblies which supply the stabilising motions; an automatic control system; and on-board sensors which monitor the satellite's position, relative to certain fixed points. Although the system normally functions autonomously, it is common practice to provide the facility for remote monitoring and over-ride by ground control via the TTC (Tracking, Telemetry and Command) system.

The Two Types of Stabilisation System

There are two types of stabilisation system; spin stabilisation and three-axis stabilisation. The choice of system is governed mainly by the shape of the satellite body, the complexity of the antennae array, and the type of solar array employed.

Spin Stabilisation

Spin stabilisation employs the gyroscopic or spinning-top principle: once set spinning about its axis, the latter's initial orientation in space will be maintained. Depending upon the satellite's size and other characteristics, the spinning speed may be anything between 30 to 100 revolutions/ minute. Spin stabilised satellites, like tops, also tend to "wobble". Steps must be taken to minimise or dampen out this wobbling or nutation motion. (A nutation damper is illustrated in Figure 5.1.)

Although the spinning motion will keep the axis of the satellite pointing in a fixed position, it will not itself ensure that the antennae maintain a constant pointing direction. The solution is to mount the antennae on a platform which rotates independently of the body around the satellite's axis, and to simultaneously rotate or "de-spin" the antennae platform in a direction opposite to the stabilising spin, thus cancelling out the effect of the latter.

Spin stabilisation is appropriate and has proved very satisfactory for cylindrically symmetrical satellites equipped with cylindrical solar arrays and relatively simple antennae configurations, But for satellites with complex multiple antennae configurations and flat solar arrays a far more sophisticated system is required. This can be readily appreciated by observing that for this class of satellite, not only have the earth pointing directions of the antennae to be controlled but the solar sails have also to simultaneously follow the motion of the sun. In these situations three-axis stabilisation is generally preferred.

Three-Axis Stabilisation

Like any other solid body, the motion of a satellite about a fixed reference point in the body can be resolved into three motions about conveniently chosen axes (as shown in Figure 5.15). The three motions are referred to as yaw, pitch and roll.

Three-axis stabilisation also employs the gyroscopic principle utilising a rotating device called a *momentum wheel* on each axis. These are driven by electric motors, the relative speeds of the motors being automatically controlled.

Apart from their stabilisation function, momentum wheels also provide a convenient way of swinging the satellite to point at a different part of the earth, thus taking the place of the on-board rocket thrust positioning system and thereby conserving valuable fuel. Acceleration or deceleration of the appropriate wheels will cause the satellite to slew in the chosen direction.

Position Measurement

Whereas the TTC system, in conjunction with the on-board satellite radio beacon, provides the information needed to measure and monitor the satellite's spatial and orbital location, on-board sensors are employed for detecting and measuring deviations in local orientation and altitude. These provide input signals to the automatic control systems which in turn supply corrective signals to the mechanical drive components, the required calculations being performed by microprocessors.

In practice, various types of sensors and external reference points are used: sun sensors, infra-red earth sensors, etc. Where very fine position control is required, sensors which take navigational fixes of stars, such as Polaris, are employed.

Positioning Accuracy

The positioning accuracy requirements vary depending upon the technical characteristics of the satellite and on the services being provided. For telecommunications services, a typical requirement is that the body axis should not deviate from the specified direction by more than ±0.15 degrees. For direct broadcast television services with their high-powered concentrated beams more stringent tolerances, such as ±0.10 degrees, are now being specified.

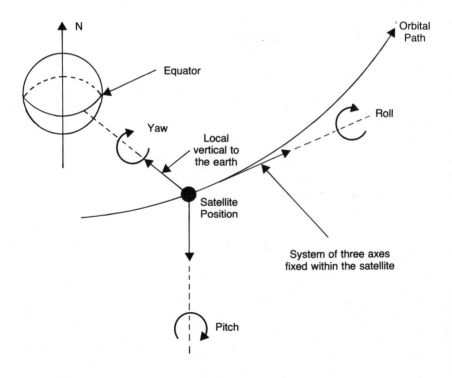

Figure 5.15   The Principles of Three-Axis Stabilisation

OTHER SUBSYSTEMS

Before concluding this overview of the principal satellite components and subsystems we briefly review two other important subsystems: tracking, telemetry and command (TTC); and the thrust subsystem.

Tracking, Telemetry and Command (TTC)

The TTC system is vitally important, both during the orbital injection and positioning phase, and subsequently throughout the satellite's operational life. The TTC system has its own dedicated radio link and its own omnidirectional antenna. The latter is essential during the launch and orbital injection phases and may continue to be used under operational conditions, although it is an increasingly common practice to transfer the TTC link to one of the transponders and the main antennae system once the satellite has passed its pre-operational tests. Because the amount of information transmitted to the ground is relatively small, only low transmission speeds are required. However, the link must be *highly reliable*.

A primary and essential function of the system is to transmit continuously a fixed frequency radio beacon generated on-board the satellite. This is activated as soon as the satellite is injected into transfer orbit. It is used by earth tracking stations for locating the satellite, and for accurately measuring its position preparatory to placing it in geostationary orbit and executing the associated positioning manoeuvres. The beacon is subsequently used throughout the satellite's operational life for monitoring its position and detecting deviations from the correct orbital position.

As well as transmitting the beacon signal, the link also transmits a wide range of telemetry data. This relates to such aspects as the performance of the satellite's subsystems and data relating to temperature and other physical conditions. All this information is transmitted back to ground control stations, where it is monitored and where appropriate commands are issued and transmitted back to initiate corrective or other actions.

The following are some of the actions which are generally available and which can be invoked by remote command:

— the thrust system can be instructed to make corrections to orbital position and altitude;

— in the event of component failure, spare components can be switched into operation;

— beam to transponder interconnections can be changed to create new communications paths between different beams;

— steerable spot beams can have their directions changed;

— manual override or back-up of on-board automatic control systems can be supplied.

Thrust Subsystem

This comprises a number of rocket motors or gas jets varying in size and power, arranged around the satellite body, together with their associated fuel tanks, distribution system and firing circuits. Both liquid and solid fuel propellants may be used and the smallest motor may be no larger than one's thumb.

Firing the appropriate motors in controlled bursts enables corrections to be made to the satellite's orbit, orbital position and altitude.

RELIABILITY

The engineering of spacecraft reliability is well understood; individual-component and overall reliability together with the various design trade-offs can be estimated mathematically. Thus, a given level of reliability, expressed in terms of probability of failure over a design lifetime of say ten years may be specified and the satellite then designed and engineered to meet this requirement. The main principles employed to achieve a high level of reliability are discussed below.

Component Selection and Testing

Only the most reliable components are used and the components and subsystems are subjected to exhaustive testing both independently and when assembled in the satellite. This requires very extensive testing facilities: special antennae test chambers and outdoor test ranges; vibration testing equipment; and special rooms that can simulate the physical conditions of space. Components which fulfil all the specified tests and performance criteria are customarily designated as "space qualified" by their suppliers.

Construction and Assembly

The satellite architecture is designed to protect components, as much as possible, from the vibrational forces encountered during launch, from the effects of thermal contraction and expansion, and from meteorite impacts in space.

Almost all subsystems, and the complete satellite, are assembled under clean-air conditions. In addition, certain critical components may be assembled in environments which are electromagnetically 'clean', to prevent spurious electromagnetism being introduced by extraneous electromagnetic fields.

Component Redundancy

Very substantial gains in reliability are obtained by incorporating spare or redundant components and subsystems. In the case of critical components (such as the receiver), there may be full duplication, whereas for items such as TWTs the ratio of number of spares to the planned operational complement might be 1 to 4 or 1 to 3.

Satellite Back-up

Finally, in order to protect against failure in operation and to guarantee continuity of service, the growing practice is to have a spare satellite in orbit and a third satellite on the ground. In the event of total failure, services are switched to the orbital spare and the one on the ground is prepared for launch.

SATELLITE LIFE-SPAN

Apart from some unpredictable event or abnormality, the life-span of a satellite is determined almost entirely by:

— exhaustion of the on-board fuel supply;

— deterioration of the electrical power supply due mainly to reducing efficiency of the solar cells and the batteries;

— general wear and tear of components.

When the fuel supply becomes exhausted, it is no longer possible to execute corrective manoeuvres, and deterioration in the power supply causes a progressive reduction in transmitted signal strength.

For some years now, the planned design lifetime for most communications satellites has been seven years and the financial justification, technical budgeting and service costs have been developed on this basis. For example, a satellite is launched with an initial power output higher than it needs, in order to compensate for progressive solar cell degradation. However, after the end of this period, there may still be sufficient power available to enable it to operate, albeit with a reduced service, for several more years. It is likely that a ten-year life-span will soon become the norm.

THE SPACE ENVIRONMENT

There are some fundamental differences between the principles and techniques of space engineering and those which have been traditionally applied in terrestrial engineering. However, although in some respects space does appear to be a very *hostile* environment, scientists and engineers now recognise that some of its properties can be turned to advantage and that for some purposes it is more suitable than earth. Apart from the deleterious effects of solar radiation and the hazards presented by meteorites and space debris, almost all the unique effects of space – and the novel engineering requirements – result from the high vacuum of space, which is more complete than can be produced on earth. The main effects of the vacuum are:

— changes in material strength;

— ineffectiveness of conventional fluid lubrication;

— material sublimation;

— absence of heat convection.

We shall first summarise the beneficial consequences before considering the adverse effects.

The Beneficial Effects

There are various effects that may be deemed beneficial:

— because components exist in the purity of an almost perfect vacuum there is no atmospheric corrosion, a continuing problem on earth;

— once the satellite launch is completed, there is no source of external vibration and the craft floats in a perfect stillness;

— there are no weather problems to contend with, although the satellite is subject to wide temperature fluctuations;

— there is no wind and almost no gravitational force, so that large, frail structures can be deployed that would be totally impracticable and certainly very costly on earth. For example, plans are now well advanced for mounting on-board the space platforms of the next decade antennae having diameters of perhaps 100 metres or more;

— because a vacuum is a good electrical insulator, electrical components can be spaced more closely together and carry higher voltages than on earth, before arcing occurs.

Effect on Materials

The mechanical properties of most materials change under high-vacuum conditions. For example, glass becomes more resistant to fracture; some metals, such as steel and molybdenum, become stronger; other metals, such as aluminium and magnesium, weaken.

Sublimation is the phenomenon whereby solid materials lose molecules in gaseous form, the molecules then being re-deposited elsewhere, such as on an insulating surface or bearing. Zinc, cadmium, lead and magnesium sublimate at lower temperatures than do many other metals and are therefore unsuitable for the surfaces of satellite components. Certain types of plastic must be avoided because some of their constituents vaporise, leaving the plastic in an extremely brittle form.

The problems of sublimation can be overcome by using suitable surface coatings and fortunately a vast selection of alternative materials suitable for space engineering has been developed. But many of these materials differ significantly from those used on earth.

Lubrication

On any satellite there are a number of mechanical joints and bearings, such as the solar sail hinges and the bearings of the antennae de-spinning system. However, the types of lubricant and lubricating principles used on earth are ineffective in space. Because of the high vacuum, fluid lubricants vaporise, and solid lubricants such as graphite lose their moisture content.

Even more serious, the traditional principles employed for metal bearings rely on the presence of a thin film of air between the bearing surfaces. But, because of the absence of air in space, metallic bearing surfaces tend to diffuse into each other and bind together through a cold welding process. The problem is solved through a combination of methods including the use of soft metals or alloys, ceramic ball bearings and bearings made of hybrid ceramic-metal compounds.

The Heat Transfer Problem

In the vacuum of space, heat convection does not take place and heat transfer can only occur through conduction and radiation. A satellite has to endure extreme changes of temperature, ranging from that of the direct sun's rays to temperatures of the order of $-150°C$ when it passes into the earth's shadow. Not only must the various components be able to function under these extremes, but the satellite's structures must be able to withstand the mechanical stresses set up by the alternate expansions and contractions.

Although a spacecraft can be prevented from becoming too hot by shielding it from direct sunlight, this solution obviously cannot be applied to the solar panels which must receive the maximum amount of sunlight. The design of the thermal control system must therefore rely on a practical compromise.

In practice a variety of temperature control and heat dissipation techniques are used including: insulating blankets and surfaces; external mirrors and other reflecting surfaces; adjustable louvres; and heat conduction pipes bonded into the structural panels. For example, a satellite might have as much as 200 or more feet of heat pipe in its external panels, in addition to reflecting surfaces and thermal insulation surfaces.

Efforts are also made to minimise the twisting and bending effects of thermal stresses on structural components. Carbon fibre materials are being used extensively for critical components such as antennae reflectors, where any significant deformation would affect both the beam properties and the pointing accuracy. Such materials are not only very light but they are also relatively insensitive to temperature changes.

Solar and Other Radiation

Although solar radiation provides the sole energy source, it also contains several forms of radiation and particles, which, together with other elementary particles emanating from outer space, have injurious effects. These comprise mainly X-rays, gamma-rays, alpha particles, protons and electrons. In fact, during its orbit, a satellite passes through some massive radiation belts consisting mainly of electrons trapped by the earth's magnetic field.

Some of the radiation such as X- and gamma-rays can penetrate deep into matter, scattering electrons, causing occasional disintegration of atomic nuclei and ionising matter. In metals it can dislodge atoms from the crystalline structure and although this generally enhances the strength of the metals, it may also lower their electrical conductivity. The latter effect could be detrimental to the performance of the on-board subsystems and their ability to handle the very tiny signal currents received from earth.

The most serious effect of electron bombardment is to degrade the performance of the solar cells. However, significant improvements in the design and protective cover of solar cells has significantly reduced the rate of degradation and this improvement may be expected to continue.

Space Debris: Natural and Man-made

It is estimated that something like 10,000 tons of meteoritic material enters the earth's atmosphere daily. The material ranges in size from tiny particles to substantial meteorites weighing 100 tons or more. The bulk of it consists of dust-like particles and most of the material is burned up in the earth's atmosphere before it reaches the ground. However, no such protection exists for a satellite and at the speeds encountered a particle measuring, for example, one millimetre across can penetrate two millimetres of aluminium.

In the early years of space exploration it was thought that meteorites would constitute a major hazard to space vehicles. However, various studies have established that large meteorites are extremely unlikely to hit a small spacecraft and that a spacecraft can survive the impact of smaller ones. For example, it has been calculated that the chance of one square metre of aluminium being penetrated to a depth of five millimetres and resulting in a hole three millimetres in diameter is once every century.

Thus, during its lifetime, a communications satellite will experience some surface erosion and perhaps a few minuscule indentations, but it is extremely unlikely that it will be hit by anything approaching the size of a walnut.

If anything, the man-made junk circulating in space presents a far greater potential hazard. According to NORAD (North American Aerospace Defence Command) there are about 5,000 pieces of redundant hardware currently in space. These include such items as nuts and bolts, discarded apogee motors, intermediate-stage rockets, and complete satellites. The latter may have been abandoned because they have reached the end of their active life, or because they failed to achieve their correct orbit during launch. Many of these defunct satellites are no longer under ground control.

These items of equipment describe a variety of orbits, some below geostationary height and others above; some either move in the geostationary orbit or intersect it. Many of the larger items are continually tracked by NORAD, but many are not, and the smaller items are, in any case, beyond the tracking capability of earth-bound equipment.

So far, there have been no orbital collisions at geostationary altitudes, and the population of redundant hardware poses only limited hazards to the 150 or so active geostationary satellites. However, a continuing uncontrolled proliferation of discarded hardware in the future could have very serious and perhaps catastrophic consequences.

There is currently no international agreement governing the effective disposal of redundant space hardware, but a number of satellite operators are taking steps to tackle the problem. For example, in order to dispose of a redundant satellite, sufficient thruster fuel is held in reserve to enable it to be pushed out of the geostationary orbit, so that it moves to a safe distance. This has now been adopted as a firm policy by INTELSAT and the European Space Agency.

6 The Transmission Link: Structure and Performance

INTRODUCTION

In this chapter we consider the radio frequency transmission link purely as a transmission vehicle. We do not discuss how it is accessed by earth stations and users; how information is impressed upon it; and those aspects of performance which have a direct impact on applications and on end-user perceptions of the service.

In the last chapter we described the principal components of the space segment transmission path, such as antennae and antennae beams, and touched upon various features of the transmission link, such as transmit/receive chains, and the possibility of interconnecting beams. It is useful to bring these various elements together and to present a consolidated picture of the transmission path to illustrate how the various features interrelate. We also consider the choice of frequencies for satellite communication, how these are allocated and co-ordinated internationally, together with the principal considerations influencing the choice of frequency.

On both the up- and down-links, the transmitted signal is subject to various losses (or *attenuation*) and its quality is also impaired through the introduction of noise. So the principal sources of loss and noise are reviewed. However, the signal also *gains* in strength through amplification at the transmit and receive antennae and in the transponder subsystem. We shall indicate how satellite communication systems designers use this information by adding up the various gains and losses to arrive at an estimate of the received signal strength.

Various other performance figures can be derived from these calcula-

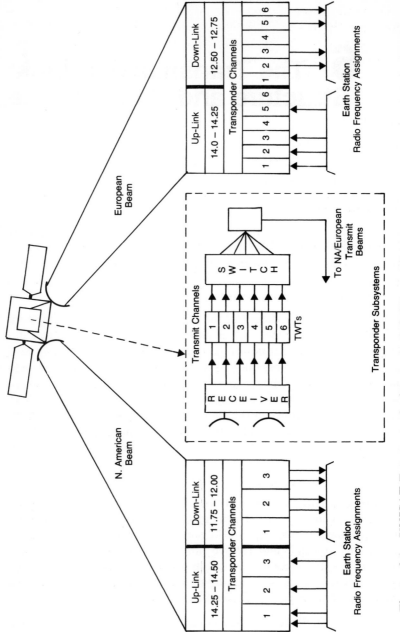

Figure 6.1   UNISAT Frequency Assignments and Channel Structure for Telecommunications Services

tions; one which is of fundamental importance is the overall *signal-to-noise ratio*. This is a major determinant of the theoretical information-carrying capacity of the link, and of the level of sensitivity required in an earth station in order that the signal can be detected and separated from the noise. We shall show, using examples, how this can be calculated and utilised.

PRINCIPLES

The general principles can perhaps be best explained by reference to Figure 6.1, which is based upon the telecommunications services section of the proposed UNISAT system. Key elements in the structure of a satellite link, considered end-to-end between two earth stations, are discussed below.

Antennae Beam Frequency Assignments

The up- and down-links corresponding to each antennae beam are allocated specific frequency bands. Earth stations in the area of coverage must operate within these bands.

Channel Subdivision

Each up-link and down-link frequency band is further subdivided into a number of channels, each channel having a frequency bandwidth equivalent to the bandwidth of an individual transponder or TWT. In the case of UNISAT the up- and down-links are allocated bandwidths of 250 MHz and the six transponders all have identical bandwidths of 36 MHz. Thus, a maximum of six channels can be catered for on each of the up- and down-links.

Transmit/Receive Chains and Path Interconnection

In order to provide a fully connected path between an up-link and a down-link, the channel structure is repeated in the transponder subsystem, the number of channels matching the number of transponders planned to be simultaneously active. Applying this to the simplest case of transmission paths which originate and terminate within the same coverage area – for example, the European beam – it can be seen that, with the full complement of transponders in operation, six independent channels can be simultaneously active.

However, the situation is somewhat more complicated when there is more than one beam, each spot beam having a different coverage area. In addition to providing connected paths within a single coverage area, there is now a further requirement to provide paths between, for example, the European up-link and the North American down-link, and perhaps vice versa. A mechanism is therefore required to ensure that the up/down-link channels and the transponder channels are connected to provide the required paths or *transmit/receive chains*.

Although these could be incorporated into the satellite on a permanent basis, this would be unduly restrictive and inflexible. Instead, the transponder subsystem is provided with a switching unit which can be activated by ground command via the TTC system. This enables the channel and transponder interconnections to be altered to give a different configuration of transmit/receive chains. With respect to a given configuration of channels, it is common to refer to a channel as being "connected" to a particular transponder.

It should be emphasised that this type of channel switching is not an instantaneous and automatic process, but can only be effected by commands from earth. Once it has taken place, the new configuration remains unchanged until receipt of further ground commands. However, a number of satellites due to come into service from the mid-1980s onwards will carry on-board microprocessor-controlled switches which will operate instantaneously and switch incoming streams of information independently of ground control. They will therefore function in a similar manner to a terrestrial telephone exchange.

It is not difficult to see that, compared with a single antennae beam, the addition of a second and subsequent spot beams multiplies the number of potential paths, the total number of possibilities generally being in excess of the number of available channels. However, the number of channels which can be simultaneously active is constrained by the number of operational transponders. In practice, a new service is usually partially justified and operated initially on the basis of what is considered to be an optimum allocation of transponders to beams, and a preferred configuration of channels and routes. Subsequently, depending upon how the service demand and traffic patterns develop, decisions may then be taken to define new paths or to otherwise re-distribute the total channel capacity.

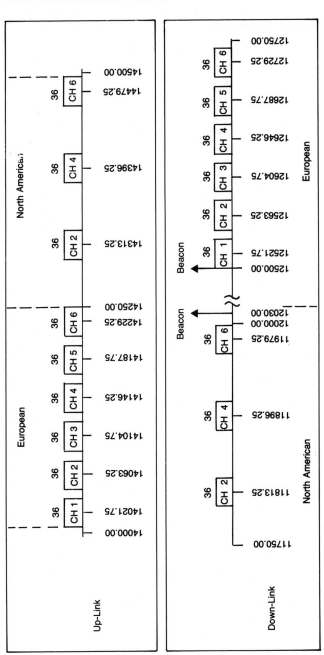

Figure 6.2   Proposed Frequency Plan for UNISAT Telecommunications Services

Notes:
1) Frequencies are given in MegaHertz ie 14250 MHz = 14·25 GHz
2) Each channel has a width of 36 MHz and is shown with its central reference frequency

Earth Station Frequency Assignments

Earth stations within the coverage area of a particular beam access one or more transponder channels assigned to that beam, and within the channels each connected earth station will operate on one or more up-link (and corresponding down-link) frequency. Whilst in general the procedures relating to the assignment of earth station frequencies will permit access either to the whole of a transponder's bandwidth or to selected portions of it, the precise details are dependent upon the particular space segment access method which is employed. These arrangements are covered in Chapter 8.

THE FREQUENCY PLAN

The information discussed above, particularly as regards frequency and channel assignments, is summarised in Figure 6.2. This depicts the proposed *frequency plan* for the UNISAT telecommunications beams, together with the central reference frequency for each channel. The down-link frequency plan also indicates the radio beacon frequencies. For technical reasons, separate beacons are required for North America and Europe.

EXAMPLE: THE ECS–1 SATELLITE

Figures 6.3 and 6.4 describe the channel structure and frequency plan for the ECS–1 satellite. It is useful to summarise features of the ECS satellites and to provide additional information to assist in understanding the diagrams. The following may be regarded as *explanatory notes:*

— the up-link and down-link frequency bands for ECS–1 are 14 -14.5 GHz and 10.95-11.7 GHz, giving bandwidths of 500 MHz and 750 MHz respectively;

— the transponder or channel bandwidth is 80 MHz. Without any frequency re-use, this would allow six channels to be accommodated;

— frequency re-use is employed over the whole bandwidth, thus permitting another six channels. In the figures the different polarisations are distinguished by X and Y symbols;

— all up-links, irrespective of the type of traffic, are received on the Eurobeam. To guard against failure, ECS–1 is equipped with a

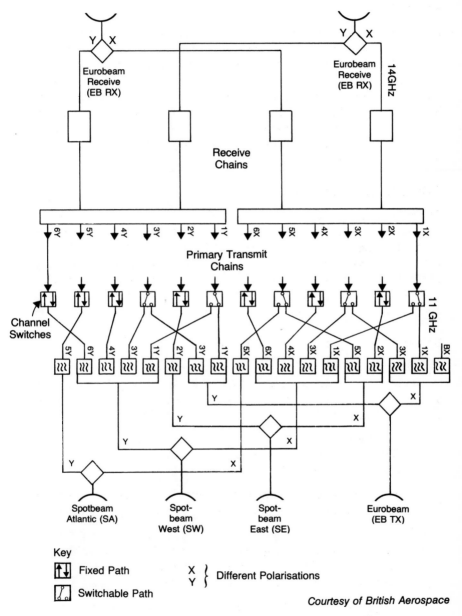

Figure 6.3   Channel Structure for the ECS-1 Satellite

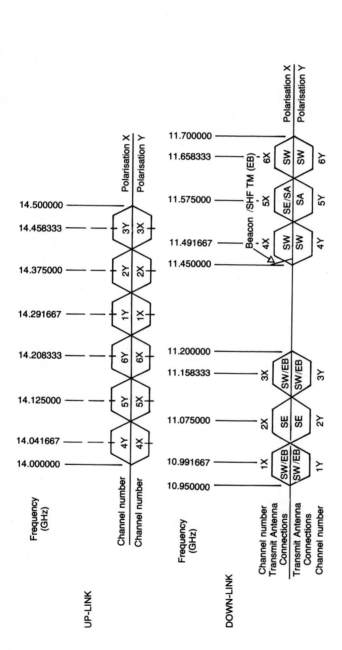

Courtesy of British Aerospace

Figure 6.4   ECS-1 Frequency Plan

spare Eurobeam antenna, and each antenna also has duplicated receivers. These are shown in Figure 6.3;

— television signals only will be transmitted through the Eurobeam, and all other telecommunications traffic will be transmitted via the Atlantic, West and East spot beams. The transponders can handle any type of traffic;

— a maximum of twelve channels is available and in normal operation it is planned that nine will be in use. Of these, five are switchable, enabling the transmission to be directed to either one of a pair of spot beams;

— Figure 6.3 illustrates the switching possibilities, and these are also summarised in Figure 6.4.

With the help of these notes and one or two examples it should not be too difficult for the reader to trace through the diagrams. For example, Channel 1X can be switched either to the Eurobeam (EB) or to the Spot Beam West (SW). Channel 2X, on the other hand, is permanently connected to Spot Beam East (SE). A separate channel (BX) is reserved for the radio beacon and the TTC channel.

FREQUENCY CONSIDERATIONS

Operating Frequency Selection

There are a number of important factors which have to be considered in selecting the operating frequency for a communication satellite.

Interference

The radio frequency spectrum is used for many purposes, in addition to satellite communications. These include: radio and television broadcasting, telephony, aircraft communications and navigational systems, mobile (ie cellular) radio, a variety of emergency services, and defence communications purposes. There are therefore a vast number of users, all contending for the useful parts of the radio spectrum, and unless the use of the spectrum is carefully controlled, they could interfere with each other, with harmful, and possibly disastrous, consequences.

Competition

The interference problem is further exacerbated by the fact that the spectrum is a scarce resource, and so some mechanism is needed to try to ensure that it is allocated in an equitable fashion.

Transmission Link Performance

The choice of transmission frequency has several important implications for overall systems performance.

First of all, as we shall see, both transmission loss and noise along the transmission link vary with frequency. As we have noted, these define the signal-to-noise ratio; this in turn determines the theoretical information-carrying capacity of the link, and has other important implications.

The antenna gain is also a function of frequency. Increasing the frequency enables smaller earth station dishes to be used. This is of major significance for earth-station costs and location, and the accessibility of satellite services.

Frequency Allocation and Co-ordination

It is evident that international agreement is required on the use and allocation of the radio-frequency spectrum for different purposes and as between different countries. This is achieved under the auspices of the International Telecommunication Union (ITU), already cited as allocating slots on the geostationary orbit. The agreements themselves originate at World Administrative Radio Conferences (WARCs) that are held from time to time. In the periods (sometimes years) between WARCs, one or more Regional Conferences may be held. These deal with specific issues relating to one or other of the three regions into which the world is divided for administrative purposes. For example a Regional WARC was held in July 1984 to allocate frequencies and orbital slots for Direct Broadcast Television Services in Region 2. The three administrative regions are:

Region 1

Europe, Africa, Asia Minor, USSR territory outside Europe, and the Mongolian People's Republic.

Region 2

This comprises most of the Western Hemisphere and includes: North and South America, Hawaii, and Greenland.

Region 3

Australia, New Zealand and those parts of Asia and the Pacific not included in Regions 1 and 2.

The ITU formalises the responsibilities of members; the agreements which are reached are contained in four documents called Administrative Regulations. Of these, the Radio Regulations and Additional Radio Regulations apply specifically to frequency allocation and satellite communications. Each of the Regulations has the force of a treaty, and once they have been ratified by a member country, the signatory is bound by International Law.

The frequency bands which are allocated internationally are in turn reallocated by the national administration and government bodies of the individual countries. The frequency band for a particular service may not be fully allocated internationally, a portion being held in reserve.

Apart from its plenipotentary conferences, WARCs and other conventions, the ITU is assisted by a number of policy-making bodies and permanent organs. These in turn are supported by various expert Committees and Working Parties. The chief ones, with their areas of responsibility, are:

CCITT (International Telegraph and Telephone Consultative Committee)
Study of technical, operating and tariff questions relating to telegraphy and telephony, and the issue of recommendations.

IFRB (International Frequency Registration Board)
Recording and notification of frequency assignments and orbital slot positions for geostationary satellites. Recommendations on earth-station requirements relating to different services.

CCIR (International Radio Consultative Committee)
Study of technical and operating questions relating to radio communications, and the issue of recommendations.

The IFRB maintains a master register of frequency assignments, and also records orbital satellite positions. It receives notification of new frequency assignments and proposed orbital slot positions, and checks them for conformity with the regulations. Through reference to the master register and the Board's weekly circular, operators throughout the

Subject	Frequency Range and Relevant Regulation	Summary
Pointing of transmitting antennae of Fixed and Mobile (Terrestrial) stations toward the orbit in bands shared with space service up-links	1-10 GHz : 470 AA	For Fixed and Mobile Stations, with EIRPs above 35 dBW: avoid orbit by 2°
	1-10 GHz : 470 BA	Where compliance with 470AA is impracticable; reduce EIRP to 47 dBW within 0.5° of orbit, increasing 8 dB per degree in directions between 0.5 and 1.5° reaching 55 dBW at 1.5°
	10-15 GHz : 470 AB	For Fixed and Mobile Stations, with EIRPs above 45 dBW : avoid orbit by 1.5°
	Above 15 GHz : 470 B	No restriction
EIRP of Fixed and Mobile Stations	470 B	55 dBW
Power to antenna of Fixed and Mobile Stations	1-10 GHz : 470C Above 10 GHz : 470 CA	13 dBW 10 dBW
EIRP of earth stations toward the horizon in bands shared with Fixed and Mobile Services	1-15 GHz : 470 G	40 dBW/4 kHz, for angles of elevation Θ, zero or below
	1-15 GHz for space-research (deep space) earth-stations only : 470 GC	40 dBW + 3Θ dBW/4 kHz for $0 \leqslant \Theta \leqslant 5°$
	Above 15 GHz : 470 GA	A flat 55 dBW/4 kHz toward the horizon
		64 dBW/1 MHz, for Θ, zero or below
	Above 15 GHz for space-research (deep space) earth stations only: 470 GD	63 + 3Θ dBW/1 MHz for $0 \leqslant \Theta \leqslant 5°$
	470 H	A flat 79 dBW/1 MHz
	470 L	Toward the horizon, limits shown above (470 G, GA, GC, GD) may be exceeded by 10 dBW by agreement
Earth-station minimum transmitting antenna elevation angle		3° above the horizontal plane (lower by agreement)
Power-flux density reaching the surface of the earth from a satellite for all methods of modulation, assuming free space-propagation conditions, in space down-link bands shared with terrestrial services. (Limits shown may be exceeded by agreement.)	1670 MHz-2535 MHz : 470 NE	in dBW/m²/4 kHz −154, for $0 \leqslant \Theta < 5°$ above the horizontal plane $-154 + \dfrac{\Theta - 5}{2}$, for $5° \leqslant \Theta < 25°$ −144, for $25° \leqslant \Theta \leqslant 90°$
	2500-2690 MHz : 470 NH	in dBW/m²/4 kHz −152, for $0 \leqslant \Theta < 5°$ $-152 + 3\dfrac{(\Theta - 5)}{4}$, for $5° \leqslant \Theta < 25°$ −137, for $25° \leqslant \Theta \leqslant 90°$

Table 6.1 Radio Regulations Affecting Orbital and Spectral Utilization

world have access to an up-to-date listing of all current frequency assignments throughout the radio spectrum. It also publishes a list of slot allocations on the geostationary orbit.

As an example of the kind of topics covered by the Radio Regulations, Table 6.1 is an extract from a larger table which lists the relevant regulations affecting orbital and frequency spectrum utilisation for communication satellites. (We cannot here explore in detail the entries in this table.)

Because of the differing technical requirements, interference implications and regulatory aspects, ITU has defined more than a dozen different categories of service that can be provided by a satellite. The principal ones are:

Fixed Satellite Service (FSS)

A Fixed Satellite Service provides two-way communications between specified fixed points. The original drafters of this definition had in mind the International trunk links between the large national earth stations.

Broadcast Satellite Service (BSS)

This covers the one-way transmission of signals up to the satellite, which are then broadcast, and capable of being received at many unspecified locations. This embraces reception by community earth stations serving local distribution systems, and direct reception by domestic receivers – what is customarily referred to as DBS.

Mobile Satellite Services (MSS)

Mobile satellite services transmit to receivers which are in motion, ie, lorries, cars, ships, etc.

Choice of Frequency

The Radio Regulations divide the Frequency Spectrum up into a number of frequency bands which are assigned to the up- and down-links for the different categories of satellite service. For example, the bands which are mostly used by the Fixed Satellite Services are: 6/4 GHz, 8/7 GHz, 14/11 GHz. The bands applicable to a particular service may themselves be further allocated across the ITU World Regions, and between regional and national services. The frequency bands are not allocated for the sole use of an individual Telecommunications Administration or service operator, and are intended to be shared. Thus, a number of Fixed Satellite Services may share the same up-link or down-link frequency band. This is one reason why international co-ordination of satellite frequency selection and orbital positioning is so very important.

A satellite can use all or part of the band allocated by the Radio Regulations. The Frequency Plan divides the available frequency band into a number of channels. As we observed in the previous chapter, it is possible to increase the number of channels by reusing frequencies, by space separated spot beams, or by polarisation.

Almost all commercial communications satellites currently in operation use the 6/4 GHz band, sometimes referred to as the "C" band, but an increasing number are being designed to operate in the 14/11 GHz or "Ku" band, either exclusively or in addition to the 6/4 band. Government and military satellites commonly use the 8/7 GHz band.

The advantages of the 6/4 GHz band are: low absorption loss from rain and water vapour; and the wide range of proven and reliable components which are available.

Major drawbacks are the extensive use of these bands by terrestrial microwave systems; and increasing congestion in the geostationary orbit. There are major sources of interference, both on the ground, and in the geostationary orbit.

The 8/7 GHz bands are currently used by governments and for military purposes, and will not be discussed further here.

The 14/11 GHz bands which are now coming into use offer the following advantages:

— they overcome the congestion and terrestrial interference problems associated with the 6/4 bands;

— they enable smaller dishes to be used;

— they allow narrower and more-concentrated spot beams to be produced;

— they facilitate the location of earth stations in built-up areas.

However, as we shall shortly see, these bands are far more vulnerable than the lower frequencies to the effects of high atmosphere precipitation, sand and dust storms and the like.

The 30/20 GHz bands are just beginning to be used in Japan and experimental work is in progress in the USA and Europe. Whereas the lower frequency bands have been allocated bandwidths of 500 MHz for the up- and down-links, 30/20 GHz bands have been allocated a much wider bandwidth of 3.5 GHz.

Interference

Types

The prime sources of interference which need to be considered and guarded against are:

— interference between satellite transmission links and terrestrial microwave links, sharing the same radio frequency;

— interference between satellites sharing the same frequency, as when one satellite receives transmissions intended for the other satellite;

— interference between earth stations accessing different satellites operating in the same frequency band, so that one earth station receives transmission intended for another earth station.

In calculating the effect of interference on transmission link performance, it is generally treated as another source of noise.

Terrestrial-Satellite Link Interference

The two most significant effects are when transmission from an earth station interferes with terrestrial link reception, and where terrestrial link transmission interferes with reception at a satellite earth station. The first, which is by far the most serious, results from the high-powered earth station signal interfering with a terrestrial microwave receiver, and the second from a terrestrial microwave beam interfering with earth station reception. The interference may be caused directly through one antenna beam shining into another, or by "spill-over". Spill-over arises because an antenna cannot be designed so that it produces a single directional beam. A number of *side lobes* are always produced, and although these can be minimised, they cannot be completely eliminated. Accordingly in order to minimise interference effects when a satellite link uses the same frequencies as a terrestrial link, it is necessary to locate an earth station so that it is a safe distance from a terrestrial microwave antenna and is not located close to a microwave transmission path.

Although for several reasons the 6/4 GHz band proved to be ideal for satellite communications, the same frequencies had earlier been assigned to terrestrial microwave services. In consequence, the major conurbations and many urban areas have, over the years, become heavily criss-crossed by terrestrial microwave links. Because of this, earth stations operating at 6/4 GHz have often had to be located some distance – as much as 50-100 miles – away from the areas which they serve. This is the primary motivation for exploiting the 14/11 GHz band, since it enables earth stations to be moved closer to their ultimate users without causing interference – into the hearts of cities or onto individual business premises. And, as we have observed elsewhere, it also results in an additional benefit in the form of smaller earth stations.

Inter-Satellite and Inter Earth Station Interference

A number of factors are responsible for the increasing severity of these effects:

— congestion in the frequency spectrum;

— closer parking of satellites in the geostationary orbit;

— the growing number of satellite systems in operation where satellites share the same frequency bands and also have overlapping footprints.

Various techniques and approaches can be employed to increase the transmission capacity of the space segment. These include: the use of new frequency bands; and increasing the utilisation of existing bands through various frequency re-use techniques.

For satellites and earth stations sharing the same frequencies, the degree of interference is governed largely by the pointing accuracy of the antennae; the power and width of the antennae transmit and receive beams; the satellite orbit spacings; and the distances apart of the earth stations. Thus, by siting earth stations a safe distance apart, and employing sufficiently narrow receive beams, it is possible to ensure that the earth stations only receive their intended transmissions.

TRANSMISSION LOSSES

Because of the distance involved, the received signal strength from a satellite may be of the order of only a few billionths of a watt. The strength of a radiated signal falls off at a rate which is proportional to the square of the distance travelled, and a satellite transmission path is 500 to 800 times as long as the path between two typical terrestrial microwave repeaters.

Reduction in signal strength arises from two sources: space loss, which is caused by antennae not being 100% efficient; and losses due to absorption of the signal by the atmosphere.

Free Space Loss

When discussing antennae principles (in Chapter 5) we gave a qualitative indication of the nature of the free space loss problem when we observed that for an isotropic antenna, only the radiation falling within a cone with an angle of approximately 18° actually reached the earth; with the remainder being lost to space. We then proceeded to explain that this loss could be reduced by designing an antenna so that it had greater directionality or *gain*.

It is possible to calculate *free space loss* for the up- and down-links in terms of the transmission frequencies and the respective *gains* of the transmission and receiving antennae. The relevant formula is quoted in reference 5 in the Bibliography, and the following example gives the result of applying the formula to a typical case.

Example

Frequency band 6/4 GHz
Up-link frequency 6.175 GHz
Earth Station Antenna diameter 30 metres
Satellite Antenna diameter 1.5 metres

Applying the formula using these figures shows that the Up-Link Loss amounts to about 99.7 decibels. This means that the power transmitted is approximately 9×10^9 times the power actually received at the satellite. This is a spectacular loss, and in fact free space loss is the dominant loss on a satellite link, in relation to atmospheric losses.

From the mathematical formulae the following qualitative results can be readily deduced:

— for fixed antennae sizes, the loss is inversely proportional to the square of the frequency, ie the higher the frequency the less the loss;

— for a fixed frequency, the loss is inversely proportional to the product of the antennae aperture areas, ie the larger the antennae apertures, the lower the loss;

— for a given satellite operating at a given frequency, the loss is inversely proportional to the area of the earth station antenna.

Atmospheric Losses

The following six factors are responsible for atmospheric absorption losses:

— oxygen in a free molecular form;

— uncondensed water vapour;

— rain;

— fog and clouds;

— snow and hail;

— free electrons in the atmosphere.

Of these, the first two are relatively constant, whereas the last four vary widely, depending on weather and atmospheric conditions. A considerable body of knowledge has been accumulated about the effects of these various factors, although the quantitative relationships are by no means simple, and reveal some interesting features. Figure 6.5 plots on the same diagram the composite loss caused by electron absorption, molecular oxygen and uncondensed water vapour as a function of transmission frequency, and also of the angle of elevation of the earth station relative to the satellite.

It will be observed that the curves on the right hand side of Figure 6.5 have two peaks round about 21 GHz and 60 GHz, and these are mainly attributable to the effect of water molecules and molecular oxygen respectively. Of the other gases, atmospheric nitrogen does not have a peak, but carbon dioxide gives rise to one at 300 GHz.

When free electrons are present, the radio waves collide with them, and lose energy through absorption by the electrons. However, electron absorption is only significant at frequencies below about 600 MHz, and electron density in the ionosphere is also greatly reduced during the hours of darkness.

Figure 6.5 also shows that there are two useful frequency "windows"

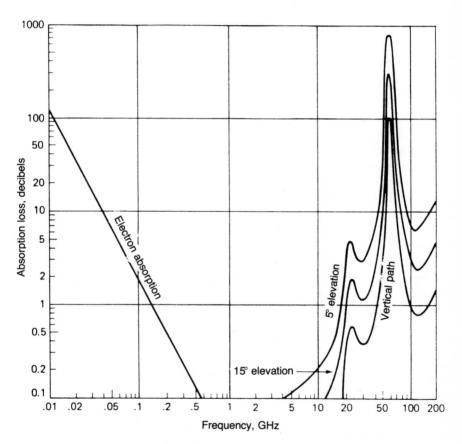

James Martin, COMMUNICATIONS SATELLITE SYSTEMS, © 1978, p 113
Adapted by permission of Prentice-Hall Inc, Englewood Cliffs, N J

Figure 6.5   Absorption Losses Caused by Electrons, Molecular Oxygen and Uncondensed Water Vapour

within which absorption effects are either negligible or have a local minimum. These occur at: 500 MHz–10 GHz and around 30 GHz between the two peaks. The first window straddles the popular 6/4 GHz band, and the second contains the 30/20 GHz frequencies which are now becoming a focus of interest. The 14/11 GHz frequencies fall in between, and although they suffer some loss, this is not excessive. Around 40 GHz, the water vapour losses increase at a rapid rate, and it is not until 90 GHz is reached that losses reduce to levels comparable with those at 30 GHz.

The effects considered above all increase in severity with reducing earth station elevation, because of the increasing length of the transmission path. The data given in the diagram applies to earth stations at sea level, and if the earth station is placed on top of a mountain, it may be possible to reduce the losses by as much as a half.

The losses discussed so far tend to remain reasonably constant and predictable, but losses caused by precipitation in the form of rain, fog, clouds, and snow and hail are very variable and far less predictable. Nevertheless, their effects can be estimated and allowed for in the design of a satellite communications system. Whilst some types of weather such as gentle rainfall typically may continue for several hours or days and affect an extended area, severe storms commonly have a short duration and affect a number of small areas in rapid succession. For these reasons, safeguarding against severe storms presents a major problem to the designers and operators of satellite communications services. In this respect the United Kingdom and a number of areas of Northern Europe are perhaps more fortunately placed than some parts of the USA such as the North Eastern seaboard and the southern states; the former is subject to heavy rainfall, and the latter experience frequent thunderstorms.

NOISE

General

Although there is a very substantial loss of signal strength on a satellite link, the received signal can be amplified to compensate for the loss. However, signal strength is not the sole determinant of the effectiveness of a satellite link. Like any other communication channel or piece of electronic circuitry there is always some noise present. This may be due simply to the random motion of electrons in the electronic circuitry and in other equipment; or it may be due to some form of external interference, such as electromagnetic radiation, weather conditions and a variety of other causes.

It follows that if the received signal is sufficiently weak in relation to the noise, it may be impossible to detect the signal at all. Furthermore, even if the signal is detectable and steps are not taken within the system to reduce the noise to an acceptable level, the noise may impair the quality of the signal. This will cause errors in the information which is being transmitted, errors which may then become apparent to the ultimate end user of the system.

Thus, neither the absolute value of the signal nor of the noise are sufficient in isolation for measuring the effectiveness of a satellite communications channel.

What is also required is a figure of merit called the *signal-to-noise ratio,* expressed symbolically as:

$$\text{Signal-to-Noise Ratio} = \frac{S}{N}$$

where S = power of the received signal

N = power of the received noise

Sometimes C, standing for *carrier,* is used in place of S; it means the same thing because the radio frequency carrier of information is the signal.

Noise can occur in two forms: it can have an essentially random pattern which is reasonably stable and always present as *background noise,* or it can occur as a sudden violent burst often with high power relative to the random noise. The former, referred to as *white or Gaussian* noise is typically present in electronic circuitry; the latter, sometimes referred to as *impulsive noise* is usually caused by some sudden external disturbance such as a violent thunderstorm.

Both types of noise have to be considered in the design and operation of satellite communications systems.

The Measurement of Noise

The power of all forms of noise on a satellite link is usually quoted in terms of *noise temperature.* It derives from the method used for measuring *thermal noise,* and arises in the following way.

The noise present in electronic circuitry is due to the random motion of electrons, and is inescapable. It is commonly referred to as thermal noise, because the electrons move faster and the noise increases with increasing temperature. Theoretical considerations show that the power (in watts) of thermal noise affecting a given range of frequencies is proportional to the absolute temperature and to the range of frequencies or bandwidth.

There are other forms of noise present on a satellite link besides thermal noise, but for some purposes satellite communications engineers find it convenient to use the same nomenclature. They do this by defining a purely imaginary temperature (equivalent noise temperature or ENT) which produces the same noise power as the various sources of both thermal and non-thermal noise considered collectively.

The power in watts of a noise source can be measured, and by using the appropriate formula for computing thermal noise, the equivalent noise temperature can be calculated.

Noise Sources

The following are the chief sources of noise in the satellite link (some atmospheric properties which reduce signal strength also contribute to noise):

—   satellite and earth station equipment;

—   the sun;

—   the moon;

—   the earth;

—   galactic noise;

—   cosmic noise;

—   sky noise;

—   atmospheric noise;

—   man-made noise.

These various noise sources have been extensively studied, and quantitative relationships developed which enable their effects to be predicted. For our purposes, it will suffice to provide brief explanatory comments.

Equipment Noise

This is contributed mainly by the antennae structures and the receiver electronics. Early earth stations used cryogenically cooled pre-amplifiers to reduce equipment noise, but with today's more powerful satellites this is no longer necessary. The stronger signal gives an acceptable signal-to-noise ratio, and this enables cheaper earth stations to be used.

The Sun

If a satellite antenna were to be pointed directly at the sun, the signal would be blotted out due to the sun's very high noise temperature.

Sky Noise

The sky has a small but measurable noise temperature. However, both the directionality and orientation of the antennae effectively protect signals from sky and sun noise.

Earth Noise

The earth contributes some background noise which is superimposed on the signal.

Galactic Noise

Galactic noise comprises radio waves emanating from radio stars in the galaxy. Its effect diminishes rapidly at higher frequencies and has a negligible effect at frequencies over 1 GHz. *Cosmic noise* refers to other stray radio noise from outer space, and this is also negligible above 1 GHz.

Atmospheric Noise

Lightning flashes are a major source of noise below 30 MHz, but have a negligible effect at satellite frequencies.

Absorption by oxygen and water vapour molecules is a significant source of noise, which is caused by re-emission of radiation following absorption. Consequently the high noise peaks correspond to the high loss peaks depicted in Figure 6.5. As for absorption it also increases with decreasing elevation.

Man-made Noise

This arises mainly from electrical equipment, and although it plagues the lower radio frequencies, it has little effect above 1 GHz. Noise caused by interference from terrestrial radio links is a separate and more serious problem (already discussed).

Severe Weather Conditions

The noise caused by heavy rain is generally greater than all other noise sources combined. Its effect is more severe at the higher frequencies above 10 GHz. Because of wide variations in weather conditions, the noise intensity can also be very variable.

COMBATTING SIGNAL LOSS AND NOISE

Because the signal-to-noise ratio is of fundamental importance to overall system performance, an important responsibility of both the designers and operators of satellite communications systems is to ensure that the ratio is maintained at an acceptable level, and that, so far as possible, there is adequate protection against signal loss and noise.

Situations in which either the signal strength and noise levels or the climatic influences are relatively stable or predictable are generally catered for in the design of the system and in the earth station specifications. Thus, for an earth station positioned at a location close to the boundary of the coverage area, or in a portion of the coverage area affected by high precipitation, the operator may specify a more sensitive earth station with a larger antenna diameter than could be considered appropriate under more favourable circumstances.

The major problems arise from the sudden storms which may blot out the signal. Although this could be partly compensated for in the earth station design, it is also likely to be an expensive solution – although much would depend upon the frequency and size of the storms. A favoured technique relies upon designing the system so that the transponders can be operated at less than full power under normal operational conditions. When adverse weather conditions arise the transponders are driven up to full power, in the expectation that the signal-to-noise ratio will increase to an acceptable level.

Noise can also introduce errors into the information which is being transmitted. A second line of defence against residual noise is to make provision for detecting and correcting errors in the received information. These functions are not performed at the level of the radio frequency transmission link, but depend upon the information flow control protocols which govern the transmission of information between ultimate end-user terminals and computers.

TRANSMISSION LINK PERFORMANCE

We have so far considered the structural components of a satellite link, and the factors that affect its performance. It is now appropriate to bring together these various aspects, and to explain how the overall performance can be determined and evaluated.

The overall performance of a satellite communications system depends upon a number of factors apart from the performance characteristics of the radio frequency link. But for the present we concentrate on the transmission link in isolation. We start by showing how the various gains and losses can be collected together and summarised in the form of the *link budget*.

The Link Budget

For any proposed system, the various gains and losses which arise can be specified or calculated and added together – the result commonly being referred to as the link budget. In effect this tells the designer what happens to the signal and the received signal strength under any given set of assumptions.

Figure 6.6, which explains the arithmetic, does not cover *all* sources of loss. It will be observed that all the quantities of interest, such as transmitter power, gains and losses, are quoted in decibels, so that they can all be added together or subtracted from one another. For example, EIRP can be computed by adding together the gain and the transmitter power, instead of multiplying them.

This simplified example concentrates on the signal strength, but by including data relating to such items as noise and received signal bandwidth, the calculations form a basis for deriving further figures of merit and other information relating to performance. Thus the received signal-to-noise ratio can be calculated, and this in turn can be used to compute the theoretical information-carrying capacity of the satellite communications channel. Tables 6.2 and 6.3 give an example of a more extensive link

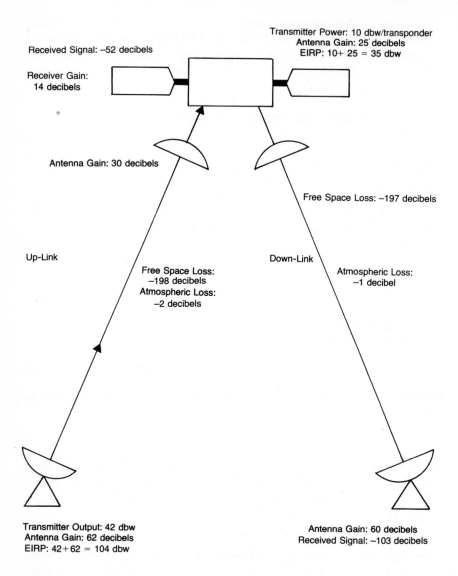

Transmitter Power: 10 dbw/transponder
Antenna Gain: 25 decibels
EIRP: 10+ 25 = 35 dbw

Received Signal: −52 decibels

Receiver Gain:
14 decibels

Antenna Gain: 30 decibels

Free Space Loss: −197 decibels

Up-Link

Free Space Loss:
−198 decibels
Atmospheric Loss:
−2 decibels

Down-Link

Atmospheric Loss:
−1 decibel

Transmitter Output: 42 dbw
Antenna Gain: 62 decibels
EIRP: 42+62 = 104 dbw

Antenna Gain: 60 decibels
Received Signal: −103 decibels

Figure 6.6   Typical Losses and Gains for a 6/4 GHz System with Large Earth Station Antennae

TRANSMITTER POWER, DBW*	25
TRANSMITTER SYSTEM LOSS, DECIBELS	−1
TRANSMITTING ANTENNA GAIN, DECIBELS	46
ATMOSPHERIC LOSS, DECIBELS	−0.5
FREE SPACE LOSS, DECIBELS	−208
RECEIVER ANTENNA GAIN, DECIBELS	46
RECEIVER SYSTEM LOSS, DECIBELS	−1
RECEIVED POWER, DBW*	−93.5
NOISE TEMPERATURE, °K	1000
RECEIVED BANDWIDTH, MHz	36
NOISE, DBW*	−128
RECEIVED SIGNAL-TO-NOISE RATIO, DECIBELS	34.5
LOSS IN BAD STORM, DECIBELS	−10
RECEIVED SIGNAL-TO-NOISE RATIO IN BAD STORM, DECIBELS	24.5

Notes:

Transmission Link : 14/11 GHz   DBW* = Decibels/Watt
Satellite Antenna : 1.8 metres
Earth Antenna : 1.8 metres
Low-Cost Earth Station Receiver

Courtesy of Satellite Business Systems

Table 6.2   Summary of Losses, Gains and Noise on a Satellite Up-Link

budget calculation. (Note: Asterisked items are quoted as decibels/watt, the power in watts being referred to one watt before converting to decibels. For example 10 watts = 10 decibels/watt, 100 watts = 20 decibels/watt. See also Appendix A.)

The reader may like to confirm that the figure given for received power in line 8 of each table, is in fact the *algebraic* sum of the preceding figures. Received bandwidth, together with a figure for noise, enables the signal-to-noise ratio to be calculated, and this appears as the third item from the bottom. In the final two items the effect of a severe storm on the signal-to-noise ratio is then evaluated.

TRANSMITTER POWER, DBw*	20
TRANSMITTER SYSTEM LOSS, DECIBELS	−1
TRANSMITTING ANTENNA GAIN, DECIBELS	44
FREE SPACE LOSS, DECIBELS	−206
ATMOSPHERIC LOSS, DECIBELS	−0.6
RECEIVER ANTENNA GAIN, DECIBELS	44
RECEIVER SYSTEM LOSS, DECIBELS	−1
RECEIVED POWER, DBw*	−100.6
NOISE TEMPERATURE, °K	1000
RECEIVED BANDWIDTH, MHz	36
NOISE, DBw*	−128
RECEIVED SIGNAL-TO-NOISE RATIO, DECIBELS	27.4
LOSS IN BAD STORM, DECIBELS	−10
RECEIVED SIGNAL-TO-NOISE RATIO IN	
BAD STORM, DECIBELS	17.4

Notes:
Transmission Link : 14/11 GHz   DBw* = Decibels/Watt
Satellite Antenna : 1.8 metres
Earth Antenna : 1.8 metres
Low-Cost Earth Station Receiver

Courtesy of Satellite Business Systems

Table 6.3   Summary of Losses, Gains and Noise on a Satellite Down-Link

In order to arrive at the optimum and cost-effective configuration, system designers perform similar calculations many times over using different satellite configurations and under a variety of assumed operational conditions, relating to weather and other environmental factors.

Information-Transmission Capacity

A fundamental equation of communications theory, due to C E Shannon, relates the information-carrying capacity of a communications channel to its frequency bandwidth and the signal-to-noise ratio of the channel:

$$I = B \log_2 \left(\frac{S}{N} + 1 \right)$$

where I = information-carrying capacity expressed in binary bits/second

B = channel bandwidth in Hertz

S/N = signal-to-noise ratio

$\log_2$ = logarithm to base 2

The equation applies generally to information in any form, whether speech, data, text, and so on. For a satellite link the channel bandwidth will be the bandwidth associated with a specified radio frequency channel which the satellite is using.

The equation enables the theoretical maximum information-carrying capacity to be calculated.

The following examples serve to illustrate the application of the above formula and to give some idea of the comparative performance of satellite and terrestrial links. For comparison purposes it is convenient to divide the theoretical information-carrying capacity (I) by the channel bandwidth (B), so that the result is expressed as information-carrying capacity/Hertz of bandwidth. The assumed signal-to-noise ratios are typical of those encountered in practice, although the satellite figure should be regarded as on the conservative side.

Example 1   Terrestrial Microwave Link

$$\frac{S}{N} = 1000 \quad (30 \text{ decibels})$$

$$\frac{I}{B} = \log_2 (1 + 1000) = 9.97$$

Example 2   Satellite Link

$$\frac{S}{N} = 10 \quad (10 \text{ decibels})$$

$$\frac{I}{B} = \log_2 (1 + 10) = 3.46$$

Limited though they are, the examples demonstrate the dramatic impact on performance of the unfavourable signal-to-noise ratios which obtain on satellite links. In fact, with current state of the art engineering, the theoretical predictions are not fully realised, as the following data shows.

For example, considering data first of all, a 36 MHz transponder can typically transmit 50 million bits/second giving a value for I/B of 1.39. In comparison, for a 3.4 KHz terrestrial voice channel operating at 9,600 bits/second, the figure is 3.2. For speech transmission (the speech being transmitted in analogue form), whereas 900 voice channels can be accommodated in 4 MHz of bandwidth on a terrestrial link, the same number of channels might require a full 36 MHz transponder on a satellite link – and for broadcast television, whilst a colour television channel needs about 6 MHz of bandwidth on a terrestrial link, it utilises the full capacity of a 36 MHz satellite transponder.

As regards data transmission across digital satellite links, it is neither easy nor particularly helpful to attempt to make broad statements about the transmission capacities which are achievable. In practice this depends upon a number of factors:

—   transponder bandwidth;

—   operating power level;

—   space segment access method, ie whether FDMA, SCPC or TDMA;

—   modulation and forward error connection techniques employed;

—   characteristics, such as the dish diameter of the earth station.

For example: a transponder accessed under TDMA may be expected to offer a capacity twice that of FDMA. And for another example: a single transponder on the ECS service is planned to have a capacity equivalent to at least 570 64 Kbits/sec data channels when accessed by a 5.5m diameter antenna. This reduces to approximately 360 channels when a 3.7m antenna is employed.

Satellite Channel Bit Error Rate (BER)

Just as the CCITT publishes bit error performance recommendations for terrestrial circuits, it also publishes similar recommendations for satellite channels.

BER is defined as the number of transmitted bits received with errors, as a proportion of the total number of bits transmitted, and is expressed in the following form: N in 1×10^x where N is commonly, but not necessarily, unity.

The current recommendations specify two alternative measures of error performance:

— the average BER over a predetermined period of time;

— the proportion of fixed-length time intervals which are either error free or experience an error rate no greater than a specified BER.

The first of these is more appropriate for voice and the second for data. However, neither is suitable on its own for all classes of service. For this and other reasons, the subject is still under study by CCITT and CCIR.

Because the channel bit error rate is a function of the weather conditions along the propagation path, it is variable and unpredictable, and can be much higher than 1 in 10^6 during heavy rain storms or heavy cloud cover. As for terrestrial services, errors may occur randomly as the result of Gaussian noise, or they may occur in bursts as the result of sudden violent storms.

The BER of the earth station to earth station satellite link could be lower than that of the terrestrial earth station access circuits, particularly if the latter employ analogue transmission and are of generally low quality. In such cases the BER of the terrestrial access circuits will need to be taken into account in determining the overall end-to-end BER.

In the US, earth station to earth station BERs equal to or lower than 1 in 10^6 for 95% of the time are typically experienced. This is a national average over a calendar year for modern all-digital satellite transmission services. However, by employing Forward Error Correction (FEC) techniques in conjunction with earth stations of appropriate sensitivity, BERs as low as 1 in 10^{-10} can be achieved.

OVERALL SYSTEM PERFORMANCE

As for terrestrial communications services, the most important technical criteria for evaluating the performance of a satellite communications system are:

— transmission capability in terms of speed and total transmission capacity;

— efficiency with which the capacity is utilised;

— error rate, measured by the ratio of transmission errors received to total information transmitted;

— reliability, measured by the frequency of service breakdown and the proportion of time that it is inoperable.

In a satellite communications system the principal components and features which determine performance are:

Satellite Characteristics

— satellite operating frequency;

— satellite transponder power and bandwidth;

— number of transponders;

— satellite antennae design;

— satellite receiver sensitivity.

Transmission Link Properties

— signal strength losses;

— noise intensity;

— signal-to-noise ratio.

Geographical and Environmental Factors

— extent of the earth coverage area;

— atmospheric and weather conditions, both predictable and unpredictable;

— terrestrial microwave interference.

Earth Station Characteristics

— earth station antenna size;

— receiver sensitivity;

— geographical location and angle of elevation in relation to the satellite.

Information Encoding and Modulation

— transmission technology: analogue or digital;

— information encoding and compression techniques;

— modulation methods.

Link Access Arrangements and Utilisation Procedures

— performance characteristics of earth station terrestrial access circuits;

— multiple access techniques, eg FDMA, TDMA;

— demand assignment capability;

— efficiency of end-to-end transmission and flow control protocols.

The majority of the topics listed under the first four headings have been discussed in this and previous chapters, and their performance implications also noted. The remaining items will receive consideration in the next three chapters.

A number of the factors and design parameters in the above list are in fact interrelated, and we have already encountered several instances of this interdependence. For example: employing a higher transmission frequency enables smaller earth station dishes to be used, but higher frequencies are also more susceptible to signal loss. On the other hand, larger antenna dishes are more sensitive than smaller ones, and so the earth station is able to receive weaker signals at low frequencies; but large dishes also cost more, and may be difficult to locate in built-up areas. And, whilst small-dish earth stations operating at higher frequencies do offer significant benefits, they also impose higher on-board satellite power requirements, resulting in a weight penalty in the space segment. These interdependencies create a whole range of trade-off opportunities, and a significant part of the task of designing a satellite communications system is to manipulate the various trade-offs, so as to arrive at a design which is near enough optimal within the constraints of cost, current technology, and service requirements.

7 The Earth Segment: Transmission, Terrestrial Infrastructure

INTRODUCTION

Chapter 6 was concerned with the transmission link purely as a vehicle for carrying information. It ignored how the information is actually carried and how the satellite link is accessed and utilised on behalf of its end users. In Chapter 2 we also indicated that the satellite link differed in important respects from a terrestrial link, in terms of its effect on different types of traffic, and on how the user perceived the service. These topics are discussed in this and the next two chapters.

In this chapter we look at various aspects of the terrestrial communications infrastructure including: the nature of the terrestrial earth station access arrangements; and encoding methods and other transformations applied to information before it is transmitted across the space segment.

These discussions and those in subsequent chapters involve some reference to the basic principles of telecommunications and data transmission. We start by reviewing the more important principles and concepts, and commenting on their role in satellite communications.

TRANSMISSION PRINCIPLES

Modulation and Modulation Techniques

Just as human speech quickly loses strength and soon becomes inaudible over all but short distances, so too would the electrical signal representing speech if it were to be directly transmitted by underground cables or by radio waves. This signal *attenuation* is overcome in two ways. One approach is to amplify the signal at predetermined points along the transmission path, although this has an attendant disadvantage in that, apart from amplifying the signal, it also amplifies any noise which is present. The more powerful solution, universally employed except over very short distances, is to superimpose the information to be transmitted – whatever its form or purpose – onto a higher frequency *carrier* wave. The process is commonly referred to as "modulating the carrier".

It is in this sense that the up/down satellite links (and sub-divisions of them) function as *carriers* of information. In general the unit which performs this function is referred to as the *modulator;* for historical reasons, when the carrier is being modulated with digital data, it is generally called a *modem*. The latter term stands for *mo*dulation-*dem*odulation, implying that, in order to recover the original signal, the reverse function of demodulation is also required.

Modulation is employed extensively in the terrestrial telecommunications network and there may be several stages of modulation in the earth segment of a satellite communication system.

Modulation Principles

A variety of modulation techniques have been developed over the years, and there is continuing research into new methods. In recent years modulation technology in general, and modems in particular, have exploited microprocessor technology to reach a high level of sophistication.

Modulation techniques currently in use can, for our purposes, be classified into six categories: amplitude, frequency, phase, delta, differential and pulse code. We give an example of amplitude modulation, but focus on frequency, phase and pulse code modulation. These are in standard use in present and planned satellite communications systems.

Modulation can be applied to a signal which is either analogue or consists of discrete digital pulses. The original unmodulated signal may arise from a telephone, a computer, a television camera, or similar equipment, and is commonly referred to as the *baseband signal*. In the following discussions, with the exception of PCM, we assume that the baseband signal is of the discrete type, ie it comprises a train of pulses corresponding to a sequence of binary noughts and ones.

Figure 7.1 illustrates how the modulation of the carrier wave is effected for amplitude and frequency modulation. The designated carrier wave is sinusoidal in form, having a regularly recurring pattern, determined by a constant frequency.

Depending upon the modulation method, some defined and measurable characteristic of the baseband signal is then used to vary either the same or a different characteristic of the carrier. As a result, changes in the baseband signal are accompanied by corresponding changes in the carrier.

Modulation Techniques

Four of the techniques mentioned earlier are illustrated in Figures 7.1, 7.2 and 7.3.

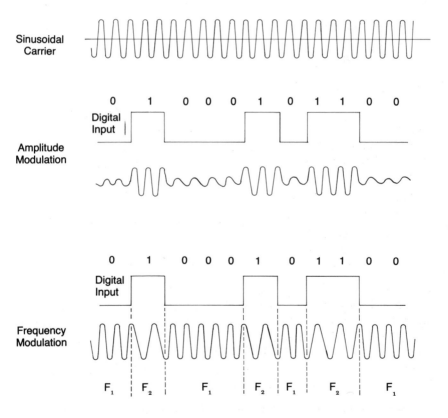

Figure 7.1   Modulation Techniques – Amplitude and Frequency Modulation

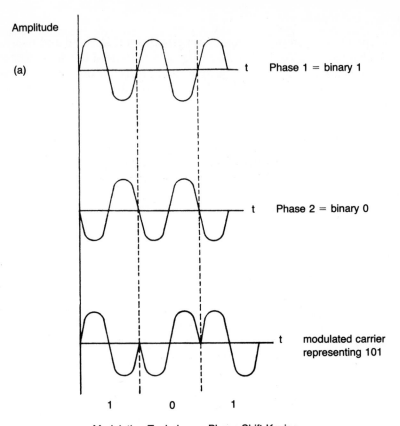

Amplitude

(a)

Phase 1 = binary 1

Phase 2 = binary 0

modulated carrier
representing 101

1   0   1

Modulation Techniques: Phase Shift Keying

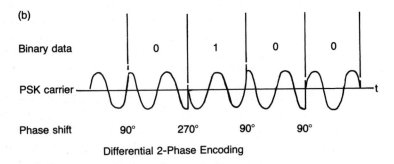

(b)

Binary data   0   1   0   0

PSK carrier

Phase shift   90°   270°   90°   90°

Differential 2-Phase Encoding

Figure 7.2   Phase Modulation

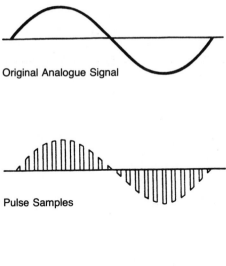

Original Analogue Signal

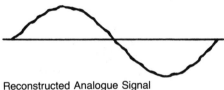

Pulse Samples

Reconstructed Analogue Signal

Figure 7.3   Pulse Code Modulation

With *amplitude modulation* (Figure 7.1) changes in the amplitude or voltage level of the signal cause corresponding changes in the amplitude of the carrier. With *frequency modulation* (Figure 7.1) changes in the signal amplitude cause changes in the frequency of the carrier, so that to represent either '0' or '1', two different frequencies are required.

The *phase modulation* technique (Figures 7.2a and 7.2b) relies upon the principle that two or more samples of the same periodic waveform (or carrier) can be arranged to have different time origins, so that they are "out of step" and are described as having different phases. In the demodulation process these phase changes can be detected and the original signal reconstructed.

Figure 7.2 shows how the sequence '101' can be modulated onto a carrier using carrier waveforms which are 180° (or half a cycle) out of phase to represent the symbols '0' and '1' respectively. The technique is commonly referred to as PSK modulation or Phase Shift Keying – by analogy with FSK (Frequency Shift Keying), another name for frequency modulation.

In the traditional telephone network, speech is transmitted as a continuously varying waveform containing frequency components in the range 300-3400 Hertz. The *Pulse Code Modulation* (PCM) technique (Figure 7.3) was not intended for transmitting data, but was devised specifically to enable speech to be transmitted in a digital form. It exploits the principle that if an analogue signal is sampled at a rate at least twice that of the highest frequency present, then the original speech signal can be reconstructed with acceptable quality from the discrete sample values.

Most PCM systems which are being installed conform to various CCITT recommendations, and sample the speech signal at a rate of 8000 times/second (ie 4000 Hertz x 2). Each sample provides a voltage level which is encoded into a 7-bit code, thus allowing any one of 32 possible values to be represented. An eighth bit is then added to represent the sign of the value. This results in a bit rate for a PCM channel of $8 \times 8000 = 64$ Kbits/second. Figure 7.3 illustrates the technique.

Other Modulation Techniques

In addition to the techniques described above, a number of advanced modulation techniques have been developed, which are now coming into service. They are intended for use in digital transmission, and are designed either to support high transmission speeds, or to enable more information to be compressed into a given carrier bandwidth.

One example is QPSK which stands for Quaternary Phase Shift Keying. This is the currently favoured method in the earth station for modulating the radio frequency carrier in digital satellite links.

We have assumed that corresponding to each change in baseband signal representing a bit, there was a corresponding change in whichever carrier characteristic has been selected for modulation. Thus, in two-phase modulation, one bit of data is encoded on the carrier for each phase change. By employing more than two phases, it would be possible to encode more than one bit per phase, and this increases the bit rate without altering the modulation rate. For example, by using four-phase changes, and coding two bits per phase change, all four combinations of two bits can be represented. The mechanics are illustrated in Figure 7.4. Modems employing QPSK are capable of generating and transmitting bit streams at burst rates of 66 Megabits/second or greater, and operate in this mode on TDMA systems. They are also operated in a continuous mode, usually at lower bit rates over digital channels on SCPC systems.

Other modulation techniques include: Delta Modulation (DM), Differential Pulse Code Modulation (DPCM), and adaptive variants of PCM, DPCM and DM (referred to by the acronyms APCM, ADPCM and ADM, respectively).

In delta modulation the input signal is sampled at a rate greater than that for PCM (typically 24 to 40 thousand times a second for good quality speech). Then, a one-bit code is used to transmit the *change in input level between samples*.

In DPCM, the difference between the actual sample and an estimate of it, based on past samples, is quantised and encoded as in ordinary PCM,

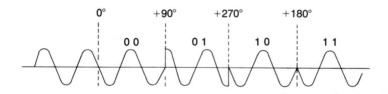

4-phase Modulation

Dibit	Phase change
00	0°
01	90°
11	180°
10	270°

Figure 7.4   4-Phase Modulation. Each Phase Change Represents a Different Combination of Binary Bit Pairs

and then transmitted. To reconstruct the original signal, the receiver must make the same prediction made by the transmitter and then add the same correction. Thus, DPCM uses the correlation between samples to obtain an acceptable performance, although it uses less bandwidth than PCM.

In adaptive PCM, DPCM and DM the quantiser has been assumed to have a fixed step size. It is possible to further improve performance at the expense of increased circuit complexity, by using an adaptive quantiser in which the step size is varied automatically in accordance with the time-varying characteristics of the input signal.

The Application of Modulation Techniques

Until recently both amplitude and frequency modulation methods were used extensively and exclusively throughout the world's telephone networks, and in terrestrial communications generally, and will continue to be used for many years to come. However, as terrestrial services are progressively converted to digital transmission, there will be increasing use of PCM for speech and phase modulation methods for data.

A similar trend is occuring in the satellite transmission link. At the present time almost all communications satellite systems employ frequency modulation, but this will change quite rapidly, as current satellites are replaced by their successors employing digital transmission.

Analogue and Digital Transmission

From the very beginning, telephone networks were designed specifically to transmit and faithfully reproduce human speech and not to transmit information in a discrete digital format. When the requirement to transmit data first arose, the development of the modem enabled information to be transmitted in this form. However, the presence of various physical components in the network continued to severely limit its digital transmission capabilities.

Beyond a certain point, the only way in which performance could be improved without adversely affecting the quality of speech transmission was to construct special circuits dedicated to data transmission.

Since the 1930s it has been known that speech could be transmitted and reproduced with acceptable quality using digital methods but, until recent years, it was neither technically nor economically feasible to employ the principle on a commercial scale. However, this did become a practical proposition with the advent of microprocessor technology, and the majority of countries and telecommunications administrations are now committed to the conversion of terrestrial transmission services to digital transmission, or are already in the process of doing so. In the UK for example, the programme is very well advanced, and by the end of the decade it is expected that all the major conurbations will be supplied by digital services. Not only will these offer vast improvements in performance over their analogue predecessors, but a digital channel is able to transmit information in all its forms: speech, data, text, facsimile, video, with equal facility. This "integrated-services" capability is of profound significance for the future progress of information technology. The following are the principal benefits offered by digital transmission:

— it is a more appropriate vehicle for linking devices which operate in a digital mode;

— it is the natural technology to employ with optical fibre cable;

— it provides transmission speeds considerably in excess of those obtainable with analogue circuits. For example, in conjunction with fibre optics, speeds of 140 Megabits/second and above are being realised along a single fibre in the cable systems now being installed;

— it allows major improvements in transmission quality as measured by induced noise and transmission errors. Noise is much reduced: on terrestrial circuits, intermediate amplifiers for increasing the signal strength can be eliminated and replaced by much simpler regenerator devices – which in effect replace the distorted and attenuated signal by a new version;

— digital transmission is more compatible with digital switching technology than is analogue transmission, and it also facilitates the use of microprocessor controlled switching.

Digital Transmission on Satellite Links

The majority of today's operational communications satellites employ analogue transmission, largely reflecting their earlier history and primary use for telephony. Analogue transmission was the only technology then available; the frequency modulation and frequency multiplexing techniques closely associated with analogue transmission were well understood. As a consequence, the methods employed for utilising and accessing the radio frequency link were geared mainly to the requirements and signal characteristics of telephony.

The use of digital transmission on satellite links offers the following advantages:

— it extends the performance improvements achieved on terrestrial circuits to the satellite link;

— with progressive digital conversion of terrestrial services, it will enable continuous end-to-end digital paths to be provided between the users of telecommunications services;

— it enables Time Division Multiple Access (TDMA) techniques to be employed for many applications, particularly those employing small-dish earth stations; this is more efficient than the traditional Frequency Division Multiple Access (FDMA) techniques, and offers a number of advantages for business use of satellite services;

— the digital format enables sophisticated signal compression techniques to be employed (see below);

— using microprocessors, it becomes possible to carry out satellite on-board processing and switching of the digital signal. As for terrestrial digital circuits, the signal can be regenerated, and this is expected to result in a two-fold increase in signal quality. The provision of a switching function will enable instantaneous switching of the receive-transmit parts to be performed.

Although the trend towards digital transmission is now well established and it will be increasingly employed on future satellites, analogue transmission will continue to be used for certain types of service and applications for some time to come.

Multiplexing

General

Multiplexing is a technique which combines the signals from a number of independent channels so that they can share a common transmission channel or other transmission facility (see Figure 7.5). At the remote end, a reverse process of demultiplexing extracts each of the original signals.

It results in major economic savings and is employed extensively in the

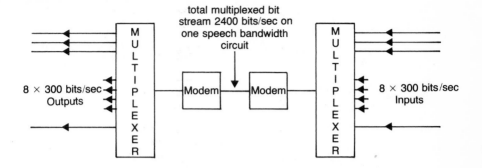

Figure 7.5   An Illustration of Multiplexing

public switched telephone network, in private communications networks, and in data communications generally.

For example, in the traditional telepnone network, although a pair of copper wires connects each telephone to the local exchange, it is more economic thereafter to group a number of speech channels together onto a wider bandwidth circuit. Similarly, in data transmission, traffic originating from 8 slow speed terminals transmitting at 300 bits/second might share a single circuit transmitting at 2400 bits/second (see Figure 7.5).

Multiplexing Methods

The techniques employed fall into three categories, according to how the capacity is shared between channels: whether by spatial subdivision, by frequency, or by time. This gives us Space Division Multiplexing, Frequency Division Multiplexing (FDM) and Time Division Multiplexing (TDM), respectively (see Figure 7.6).

The information to be transmitted can be regarded as occupying a two-dimensional continuum of frequency and time, so that the quantity of information that can be transmitted is proportional to the bandwidth and the time for which it is used.

In *Space Division Multiplexing,* a number of physical transmission circuits or paths are grouped together into one physical circuit. For example, a traditional telephone cable may contain a large number of

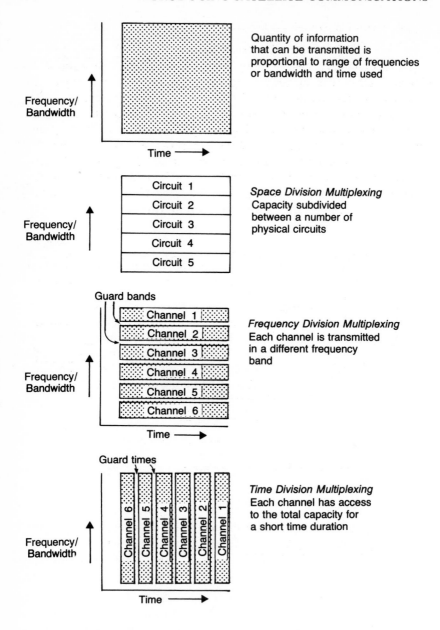

Figure 7.6   Multiplexing Techniques

copper wires. Conversely, it subdivides the *space* of a *single* cable into a number of physically independent paths.

Radio broadcasting is a familiar example of *Frequency Division Multiplexing* (FDM). The signals received by a domestic radio receiver comprise many different radio programmes which are transmitted simultaneously but which occupy different frequencies on the bandwidth of the receiver. Each radio broadcasting station is assigned its own specific carrier frequency, and the speech and other sounds that it is required to transmit are modulated onto the carrier frequency. The process of adjusting the tuning knob enables one such carrier frequency to be separated from another. The received carrier signal also has a bandwidth associated

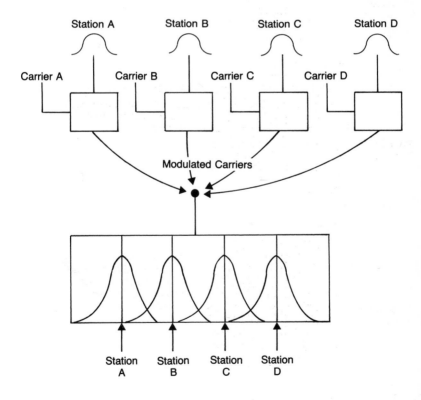

Figure 7.7   Frequency Modulation

with it, since it must be able to carry and faithfully reproduce the range of frequencies present in the transmitted programme.

The process is represented diagrammatically in Figure 7.7, which shows the received signal bandwidths centred around the station carrier frequencies. The diagram also illustrates an important property of FDM, which cannot be entirely eliminated, and that is the overlapping of adjacent frequency bandwidths. If this is excessive, it can cause *intermodulation* between adjacent carriers, resulting in distortion and interference. In a similar way, a carrier which has greater power than its neighbour can interfere with or drown out the latter.

For these reasons, in FDM adjacent channels are separated by guard bands. However, their protection is not obtained without some cost, since the guard bands reduce the total useful transmission capacity, and the greater the number of channels into which a given bandwidth is divided, the greater is the loss.

One other feature of FDM, which distinguishes it from TDM, is that the capacity of each multiplexed channel is restricted to the capacity of the subdivision of total bandwidth which is allocated to it.

Like frequency modulation, FDM fits naturally into the world of analogue transmission, whereas Phase Modulation, PCM and Time Division Multiplexing are appropriate to digital transmission.

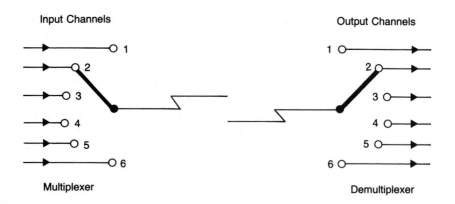

Figure 7.8   The Time Division Multiplexing Concept

In *Time Division Multiplexing* (TDM) each multiplexed channel is connected in turn to a high-capacity channel for a short time duration. Conceptually, a Time Division Multiplexer can be viewed as a multiple position switch which connects each channel in turn, 'round-robin' fashion, to the shared resource (see Figure 7.8).

Unlike FDM, in Time Division Multiplexing, each input channel has sole use of the total channel capacity, albeit for a limited period. For this reason, it is particularly well adapted to transmitting information at high speed and in short bursts. In order to apply TDM, information must be encoded and transmitted in digital form.

Hierarchial Modulation and Multiplexing Schemes

In practice, communications channels typically involve several modulation and multiplexing stages between their origin and termination. Figure 7.9 illustrates the principle and the multiplexing scheme which is employed in a traditional analogue public switched telephone network.

Individual telephone channels are combined into five groups, each of 12 channels. Each group which occupies a frequency band of 60-108KHz is then used to modulate separate carriers to form 12 channel groups. The five groups are then combined into a supergroup which occupies a frequency band of 312-552KHz. (A supergroup is modulated onto a carrier frequency of 564KHz.) A further stage of multiplexing and modulation may be applied, resulting in 15 super groups being combined into a master group comprising 900 telephone channels.

The new digital transmission networks are constructed according to similar architectural principles. For example, in the UK, thirty-two 64 Kbits/second channels are combined into a 2.048 Mbits/second channel and through further multiplexing stages up to 139.264 Mbits/second. The 2 Megabit channel carries 30 speech channels plus two channels used for signalling and control purposes. In North America and Japan however, the situation is slightly different, and follows an alternative CCITT recommendation. Instead of 64 Kbits/second, a speed of 56 Kbits/second is used, 24 such channels being combined into a 1.544 Megabit channel. For comparison, these are shown in Table 7.1.

Multiplexing in Satellite Communications

The three types of multiplexing so extensively employed in terrestrial communications are, in conjunction with modulation, also important in

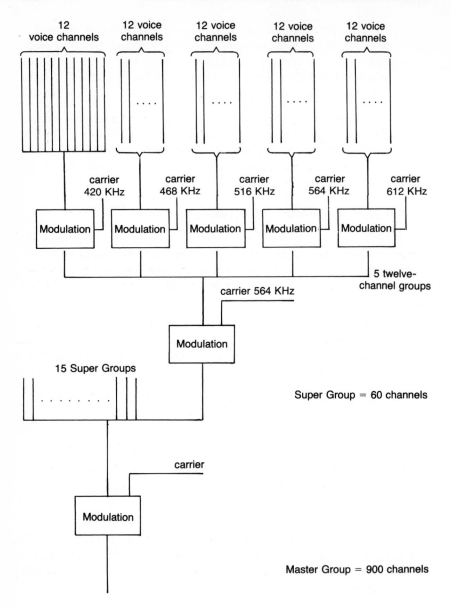

Figure 7.9   A Typical Multiplexing Hierarchy for a National Telephone Network

North America and Japan	UK and Europe
56 Kbits/sec	64 Kbits/sec
1.544 Mbits/sec	2.048 Mbits/sec
3.152 Mbits/sec	8.448 Mbits/sec
6.312 Mbits/sec	34.368 Mbits/sec
44.736 Mbits/sec	139.264 Mbits/sec
274.176 Mbits/sec	

Table 7.1   The CCITT Digital Network Hierarchies

satellite communications. The principles are utilised, either directly or indirectly, at various stages along the transmission path: between the user and the earth station; within the earth station; and in the satellite link access procedures.

Multiplexing may be necessary to combine traffic from a variety of sources onto the access circuit connecting the user with the earth station. Sources include: computing equipment, terminals, PABXs, and Local Area Networks (LANs).

The access circuit to an earth station remote from the customer's premises will have been derived from the speed hierarchy of the terrestrial network.

Depending upon the design of the earth station and the service characteristics, an initial multiplexing stage may be present, and at least one stage of modulation is certainly necessary.

Space Segment Access Arrangements

In Chapter 5, attention was given to how satellite capacity could be simultaneously shared by a number of channels by employing separate spot beams and frequency re-use. This can also be regarded as an application of Space Division Multiplexing.

Finally, Frequency Division and Time Division Multiplexing principles also provide the basis for the two principal space segment multiple-access systems, FDMA and TDMA respectively.

TERRESTRIAL TRANSMISSION PATH AND BASEBAND SIGNAL PROCESSING

Figure 7.10 shows a schematic of the terrestrial link between a user's premises and an earth station, the latter being located either on the user's premises or remotely situated. In the former case the link will be by a circuit located on the user's premises, whilst in the latter it will be derived from the terrestrial network, in the form of either a cable or a microwave link.

In broad functional terms the transmit/receive path in an earth station can be subdivided into three stages in both the up-link and the down-link directions: baseband, Intermediate Frequency (IF) and Radio Frequency (RF).

The baseband signals generated or received by the end user's equipment are subjected to a variety of processes. These may occur on the customer's premises, they may be inherent in the terrestrial transmission link, or they may be required just for transmission over the satellite link. Whatever the reason, the baseband processes substantially influence the

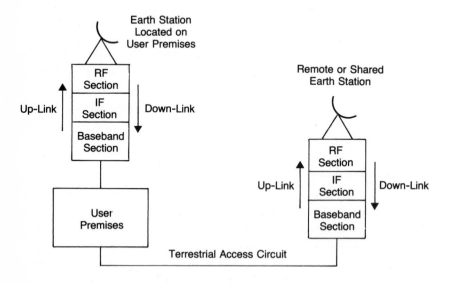

Figure 7.10   Earth Station Access

performance of the satellite link. These are the principal operations:

— modulation;

— multiplexing;

— analogue-to-digital conversion and signal compression;

— voice activation;

— echo suppression or cancellation;

— speech interpolation;

— encryption;

— forward error correction;

— delay compensation.

Modulation and multiplexing have been covered in sufficient detail for the purposes of this book.

Voice activation is a method of achieving a bandwidth/power advantage through a device that withdraws the transmission channel from the satellite network during pauses in speech. It has found a practical application in SCPC systems such as SPADE (but see Chapter 8). Echo suppression and delay compensation are more appropriately considered in Chapter 9 but we shall consider the remaining ones here. Techniques such as speech interpretation and signal compression are employed in order to improve link utilisation by compressing more information into a given bandwidth, whereas the purpose of encryption is to ensure that information is unintelligible to an unauthorised recipient.

Analogue-to-Digital Conversion

Although the trend towards digital transmission is now firmly established, it has suggested some interesting implications. One is the relative efficiency and cost-effectiveness of digital, compared with analogue, transmission, in respect of the principal types of traffic, ie data, voice or video.

In terms of costs and capabilities, traditional analogue transmission services have favoured voice; data transmission being relatively more costly, and also subject to speed limitations. With digital transmission the situation is almost reversed. Whilst high bit rates can be readily supplied at relatively low cost, the 64 Kbits/sec required for a single speed channel could be considered expensive in terms of the transmission bandwidth that it consumes. Accordingly, in recent years a significant amount of research has been directed towards improving the efficiency of digital transmission for information other than data.

The efficiency of digital signalling depends upon two factors: firstly, on the transmission speed itself, and as we have seen this is partly dependent upon the modulation techniques and the capabilities of the modem; and secondly, on how many bits/sec are required to digitise the human voice, music, television, or whatever else is transmitted. By applying the theorem referred to when discussing PCM, the theoretical bit rates for various types of information can be calculated, and these are reproduced in Table 7.2.

Since the 64 Kbits/sec (or 56 Kbits/sec alternative) recommendations governing the digital representation of speech were drawn up by CCITT, there has been remarkable progress in the development of techniques for compressing information into fewer signals, and enabling lower transmission speeds to be used. Although satellite communications provided a major stimulus, these techniques are also being applied in the new terrestrial digital services.

Type of Information	Bit Rate
Fast Facsimile	64 Kbits/sec
Slow-Scan TV	64 Kbits/sec
Speech (Telephony)	64 Kbits/sec
High-Fidelity Sound	250 Kbits/sec
Picture Prestel	1 Mbits/sec
TV 625 Lines Colour	68 Mbits/sec
High-Definition TV	144 Mbits/sec

Table 7.2   Bit Transmission Rates for Different Types of Information

Although the principles have wide applicability, here we shall consider their application to speech and to video conferencing.

Digital Speech Compression

Signal compression in voice channels is achieved primarily through the application of advanced modulation techniques such as differential PCM and delta modulation combined with adaptive algorithms. Instead of encoding *absolute* signal values, these modulation techniques rely upon detecting and encoding differences between successive signals, or between a signal and its predicted value, the prediction being based upon previous samples. The incorporation of an adaptive algorithm enables the sampling circuits in the modem and the number of encoding bits/sample to adapt to the changing characteristics of the particular section of speech.

Speech can now be transmitted digitally and reconstructed to acceptable quality at speeds of 32 Kbits/sec and 16 Kbits/sec (32 Kbits/sec is now in common use on satellite links). Although there is continuing research to achieve even greater compression, 10-12 Kbits/sec seems to be regarded as the practical lower limit.

Video Compression

Table 7.2 highlights the high digital transmission speed required to faithfully reproduce full motion pictures to the same standard as broadcast TV. (Full motion refers to an image which reproduces a wide range of movement without detectable blurring.)

The growing interest in the potential benefits of video conferencing have provided a major stimulus for the development of video compression techniques. As a result, codecs (the devices which digitise speech or video signals) are now available which compress full colour video signals into either 2 or 1.5 Mbits/sec. Codecs employ various sophisticated algorithms which monitor the signals, and in conjunction with adaptive prediction reduce the number of bits required to represent the image. For example, one such algorithm operates by arranging for only those parts of the information that have changed significantly to be transmitted.

Adaptive Differential Pulse Code Modulation (ADPCM) is now the preferred modulation technique for video image digitisation, and is employed in the codecs now becoming available.

The resulting screen images are perfectly satisfactory for video conferencing and similar applications. For all but fast movements, they can reproduce motion without significant blurring.

Digital Speech Interpolation (DSI)

Speech Interpolation is a technique for improving the utilisation of telephony channels. Instead of compressing or reducing the number of signals, it makes use of the idle transmission time resulting from natural pauses in a telephone conversation. Extensive studies of telephone traffic have shown that, on average over a group of circuits, the actual level of traffic activity is substantially less than the design capacity.

During a conversation both parties do not normally speak at once, and for a small part of the total connection time (around 10%), during the natural pauses in the conversation, nobody is speaking. Furthermore, long-distance channels are carried by a four-wire circuit which provides a transmission path in both directions simultaneously. Thus a single one-way path is in use for approximately 45% of the time. Or, stated another way, for every 100 speakers, only about 45 will be speaking simultaneously.

In 1959, in order to improve the utilisation of the transatlantic telephone cable, Bell Laboratories developed a technique – Time Assignment Speech Interpolation or TASI – which effectively doubled the available capacity of the cable. TASI is a high-speed switching system that maximises the utilisation of the idle periods in a two-way telephone conversation. The TASI equipment monitors each user's speech, and assigns him a channel when he begins to speak. The speaker retains the channel until he stops speaking when it is then requisitioned for another conversation. Equipment at each end of the channel reassembles all the different conversation bursts into coherent speech. This all happens without any perceptible loss of speech quality and the user is unaware of the switching operations. TASI effectively doubles the useful capacity of the transoceanic cables where it is employed.

With digitised speech, a digital version of TASI – Digital Speech Interpolation (DSI) – is employed. This is much faster and somewhat more efficient than TASI, its analogue precursor. DSI is employed extensively on satellite links where it increases the capacity of telephony channels by a factor of 2.3 or higher. It will also be used in place of TASI on the new optical fibre transatlantic cables.

Forward Error Correction (FEC)

Forward Error Correction is a technique for detecting and correcting transmission errors by the transmission link. Because a transmission vehicle cannot be assumed to be perfectly reliable, error detection methods are an essential requirement in any data transmission system.

All error detection methods depend upon subdividing the information into blocks or messages, and attaching to each block one or more additional bits which are a function of the information in the block. The additional information or checksum may be a single 'parity' bit, or a more complicated number. The checking information is calculated before transmission, and is recalculated at the receiving end. Any discrepancy between the transmitted and recalculated checksums indicates that a transmission error has occurred, whereupon the receiver requests a retransmission, with the expectation that it will be received correctly.

These are the underlying principles of the error detection techniques employed in terrestrial data transmission systems for many years. However, in order to compensate for the poor signal-to-noise ratio and to maximise throughput, much more powerful techniques may be required, at least over the space link. To rely solely upon terrestrial techniques could result in great inefficiencies, particularly at high transmission speeds.

Forward error correction differs from the technique described above in that it not only detects errors, but it also corrects them. Thus, a '0' bit which should be a '1' bit is changed to a '1', and vice versa. FEC principles were originally developed and applied to the construction of systematic binary codes for representing information. The International Alphabet number 5 is an example of a systematic code.

Theoretical studies showed that by adding a sufficient number of check bits to the information bits, and arranging the bits in appropriate patterns, a general class of codes could be constructed with the error detection and correction property. These are called "Hamming codes" after their inventor. They have rarely, if ever, been employed in practice on terrestrial circuits mainly because they entailed the transmission of significant quantities of redundant information, and in consequence were generally inefficient at the relatively low terrestrial transmission speeds. The application of FEC to satellite links is conceptually similar, but the mechanics are rather different. Instead of being based on the use of systematic codes, the procedure operates continuously on the bit stream, employing a technique called Viterbi Convolutional Encoding, the hardware component being designated a Viterbi Convolutional Encoder. Different levels of correction can be achieved, and conventionally these are identified by the following notation: 3/4, 7/8, 1/2. These signify the ratio between the number of input bits to the number of output bits, so that for FEC 1/2, the total number of output bits is approximately twice the number of input information bits. The FEC function is invariably performed in the earth station.

In some satellite systems FEC is not applied permanently, but is invoked at the operator's discretion in order to compensate for a temporary degradation in performance, which might for instance occur under sudden storm conditions. By contrast, it will be a permanent feature on the business services transponders of the European ECS–2 satellite, because, in terms of the signal-to-noise ratio, the transponders are operating very close to their acceptable limits.

Encryption

Like terrestrial radio, satellite communications is an inherently broadcast medium and transmissions are highly vulnerable to interception by other than their intended recipients. Any earth station within the same coverage area of the satellite receives the transmission of other earth stations, each earth station filtering out or discarding those transmissions which are not intended for it. The problem is particularly acute where earth stations access the same transponders, and where the service operating plan requires each earth station to be capable of transmitting or receiving throughout the full frequency bandwidth of the transponder. For example, earth stations accessing the business services or Satellite Multiple Services (SMS) transponders on the European ECS satellites commencing with ECS–2 will be configured in this way. Accordingly, systems which offer data transmission or business services provide encryption and decryption facilities in the earth stations, and these are generally available as a user option.

However, when encryption is provided at an earth station, it only operates across the space segment, and does not extend across the terrestrial access links, between the user and the earth station. To obtain full end-to-end encryption, users would need to implement their own procedures.

The ECS–2 business services will offer encryption facilities using a technique based upon the DES (Data Encryption Standard). Like DES, it is a single-key method but utilises a different algorithm. Eventually, single-key systems will be replaced by the more powerful public-key system.

Until recently, attention has been mainly focused on the encryption requirements for data and text. However, both speech and video traffic are also equally vulnerable to unauthorised reception, and the anticipated growth of satellite broadcast television and video conferencing is prompting the extension of encryption techniques to these applications.

Television distribution by satellite has for some time been a major industry in the United States, and it is now in the process of being rapidly introduced in other parts of the world. In America the "private" reception of television transmissions has become a major problem to the programme and film makers, to the broadcasting companies themselves and also to their cable distribution affiliates. Unauthorised reception of pay TV programmes is widespread: according to one industry source over 200,000 home earth-station receivers have been sold and many homes, hotels, etc are by-passing the cable systems. It is also known that some cable distributors are also deliberately distributing programme material without authorisation and payment. The emergence of the new DBS services, accompanied by anticipated reductions in home receiver costs, will obviously exacerbate the problem.

It is therefore not surprising that the TV industry should seek to protect their revenues and investments; and this is now happening, following the lead taken by the HBO (Home Box Office) organisation. HBO is the dominant, and currently most profitable, US pay TV company, transmitting to more than 5,000 cable systems across the country. In 1982, it announced a multi-million dollar commitment to introduce an encryption system. Cable TV is particularly vulnerable because the signal decoders, are, so to speak, "in the hands of the enemy" and are available in the consumer market place. The security procedure cannot therefore be entrusted solely to the remote decoder.

In the systems which are currently under development or have reached the market, a common approach is to scramble the audio and video signals at some central originating point before they are transmitted over the satellite link, and then to reconstitute the signals in a decoder unit either at the cable head end, or on domestic premises in the case of DBS.

The decryption operation is rendered immune from tampering by arranging for the key to be transmitted to the decoder on top of the main signal, and arranging for it to be changed at frequent intervals.

The satellite TV experience has obvious lessons for video conferencing. Although it has not yet come into widespread use, video conferencing is susceptible to the same abuses as broadcast TV, and unauthorised access could pose a major business threat. Unless steps are taken to scramble the transmissions, it will for example be possible for an eavesdropper to listen in to another organisation's discussions on corporate strategy, and to gain access to a competitor's sales figures and other financial information. Apart from internal small company meetings, video conferencing is being adopted increasingly for large audience conferences, and the trend is expected to continue. Without effective control over access, it would therefore be possible for people to "attend" such a conference without actually paying a subscription fee.

To date, encryption requirements have been concentrated on the transmission link and the protection of commercial information, but it is now recognised that the satellite itself is vulnerable to external interference through unauthorised access to its Tracking, Telemetry and Command (TTC) link. All that is needed is access to the proper equipment, and a prankster or even a foreign power could send the satellite out of orbit, causing heavy financial loss and untold chaos to communications services. Such aspects are under active study, and it is expected that satellite command links will soon be encrypted.

TERRESTRIAL ACCESS ARRANGEMENTS

A full review of current terrestrial access arrangements, both worldwide and on a country basis, would reveal a somewhat confusing picture. This is due to several factors, for example:

— the historical role of INTELSAT;

— the impact of technology and the emergence of lower-cost earth terminals in particular. (The term earth terminal is now coming into general use, and seems set to displace earth station so that from now onwards we use them interchangeably.);

— competitive pressures, both domestic and international;

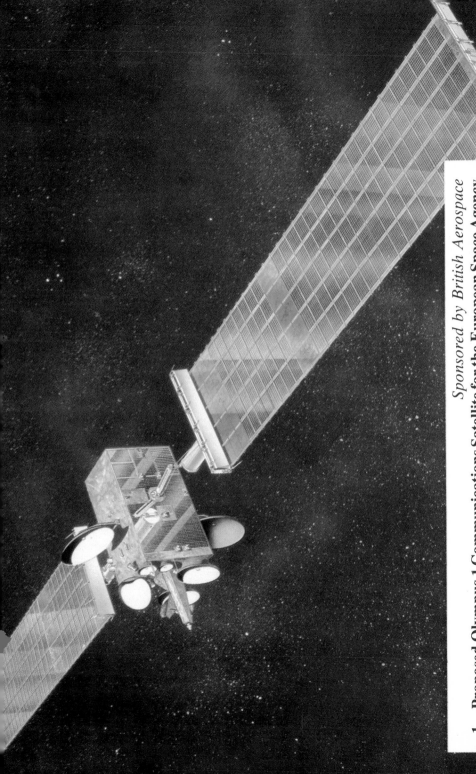

Sponsored by British Aerospace

1 Proposed Olympus 1 Communications Satellite for the European Space Agency

Sponsored by British Telecom International ⬡BTI⬡
2   SatStream Small-Dish Earth Terminal in London

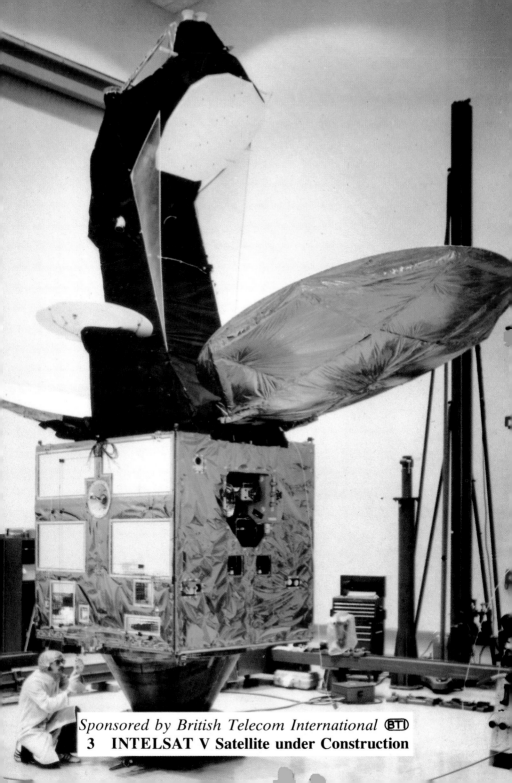

Sponsored by British Telecom International ⒷⓉⒾ
3 INTELSAT V Satellite under Construction

Sponsored by Marconi

4   30-metre British Telecom Receivers at Madley Earth Station

Sponsored by British Telecom International (BTI)

5   BTI's Transportable SatStream Earth Terminal

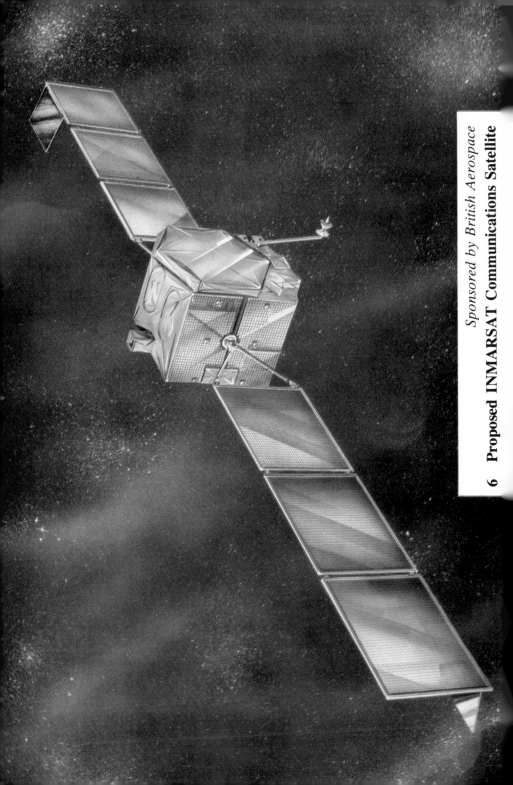

Sponsored by British Aerospace

6 Proposed INMARSAT Communications Satellite

7   11-metre Standard B Earth Station (Falklands)

Sponsored by Marconi

DANGER
MOVING
STRUCTURE

Sponsored by Marconi
8   19-metre Standard C Earth Station (Goonhilly)

Sponsored by British Telecom International (BTI)

**9   BTI's 5.5-metre Offset Gregorian Terminal in London,
offering SatStream Services to North America**

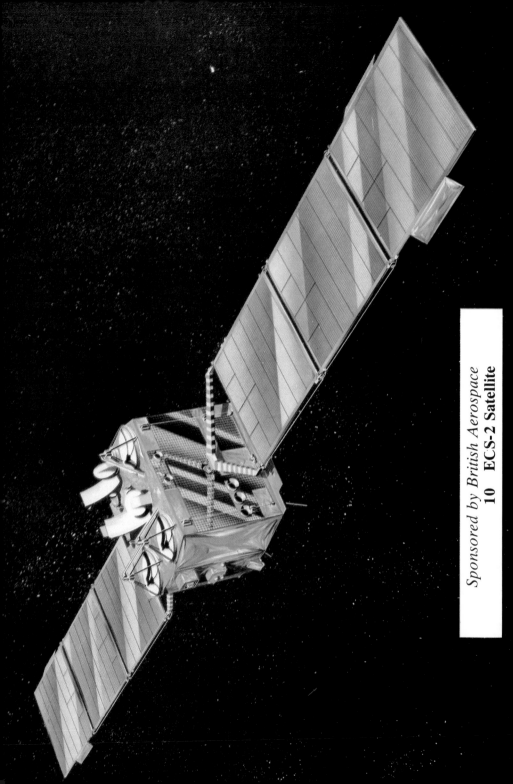

10   ECS-2 Satellite

Sponsored by British Aerospace

11 Unmanned Space Platform

Sponsored by Marconi

12   14.5-metre INMARSAT Receiver (Goonhilly)

— different national policies towards the supply of telecommunications service and the level of permitted competition.

For example, for some time now, INTELSAT has been supplying domestic satellite services to various countries. On the other hand, unable to prevent countries from establishing independent regional and domestic satellite services, it has leased the capacity and then re-leased to the countries concerned. Now, faced, on international bodies, with growing pressure for greater competition, INTELSAT has recently introduced specialised business services.

At the national level, there are wide differences between countries in respect of the policies governing the supply and operation of telecommunications services. Some countries (eg West Germany) exercise a tight monopoly, whereas others (eg the US) allow considerable competition. The United Kingdom is now embarked on a path of controlled liberalisation; there has already been considerable relaxation with regard to both the supply and utilisation of products and services. The overall situation remains very fluid and will continue to evolve. (These issues are further discussed in Chapter 10.)

Shared and Private Earth Terminals

Now that it is practicable to locate relatively low-cost earth terminals on, or adjacent to, customers' premises, thus bypassing terrestrial circuits, the question arises of the relative economies of the two forms of access. No simple and universally valid answer is possible: much depends upon circumstances and a number of interrelated factors (discussed below).

Corporate Requirements

Obviously, an organisation must have the applications and generate and receive the traffic to warrant the cost.

Terrestrial Circuit Access Costs

Irrespective of traffic and application considerations, in a large land area with a low earth station density, the cost of the terrestrial access circuits to the nearest earth station may be prohibitive. In this case an under-utilised private terminal may prove economic.

Adequacy and Penetration of Terrestrial Services

Terrestrial services may be inadequate on two counts: they may not match the performance capabilities of the space segment, and therefore constitute the weak link in the end-to-end path; or, terrestrial services may be either non-existent or rudimentary. In the first case, the application may justify an on-site earth station; in the second case, practical necessity would dictate it.

Thus, the following issues are central to the debate: adequacy and stage of development of terrestrial services; density of demand in relation to the geographical area; and the cost differentials between the different modes of access.

Since cost insensitivity with distance and geographical "reach" are amongst the major benefits of satellite communications, it is sensible to suppose that the arguments would be weighted differently for a smaller country than for a larger one. The United Kingdom in particular has a very much smaller land area than the US, and possesses a terrestrial network with very significant penetration, which is at an advanced stage of modernisation. Moreover, apart from the principal multinational companies, a large section of British industry and commerce comprises small-to-medium-sized organisations, unlikely to be handling significant amounts of information. This means that, at least in the short term, there is a case for substantial reliance on shared earth stations.

The User Interface

The requirements at the user's location vary widely depending upon how the organisation intends to utilise a satellite service. At the simplest level, a telephone suffices. It is likely that any particular international telephone call will be routed via the INTELSAT system, as has been the case for some years.

For someone wishing to transmit data, the minimum requirement is a suitable modem to connect a terminal or computer to a terrestrial circuit. If it is required to connect a number of terminals, then cost considerations would favour the addition of a multiplexer to concentrate traffic onto a single access circuit.

For a user who wants to establish a private speech circuit or network, additional equipment may be needed, depending upon such factors as:

the extent to which digital technology is employed both internally and on the access circuit and space segment, and also on the space segment Multiple Access method. For example a codec would be required to convert to PCM.

For local access to video conferencing facilities, a video compression codec together with the terminal is a minimum requirement for desk to desk or a small meetings situation. A studio-quality environment demands additional specialised equipment depending upon the facilities to be provided, and devices which perform various ancillary functions may also be required. In addition to encryption, some form of echo suppression will be needed for private speech channels, and, if propagation delay is a serious problem for data channels, then a *delay compensation* or *protocol conversion unit* would also be needed.

The acquisition of a private earth terminal constitutes a significant expansion of on-site hardware. Various facilities would have to be provided which are shared in a suitable earth station. These include the transmission electronics, control equipment, and facilities for encryption and forward error correction.

For an organisation employing satellite services for a variety of different types of traffic and applications, a substantial local communications infrastructure is required to link everything together and to feed information to the earth-terminal access circuit. (These considerations relate to such topics as local area networks, multifunction PABXs and the like.)

Encouraged by the move towards digital transmission, traffic integration is now the universal goal. However, despite the technical and economic benefits, there are circumstances in which channels carrying different types of traffic should be kept separate, at least for some portion of their common path. For example, echo suppression cannot currently be applied to data channels. This is because of "speech clipping" which can be tolerated in speech, but has adverse effects on data. And, since speech encryption and data encryption employ different techniques, they must also be handled separately.

Terrestrial Network Access

Transmission services are developing rapidly. In particular, both terrestrial and satellite services are being converted from analogue to digital technology. However, the conversion of satellite links with global coverage will be completed long before the task is completed on the ground. Therefore traditional analogue transmission systems will remain for some time, though digital terrestrial circuits will be used to an increasing extent. The rate at which continuous digital end-to-end paths become available will depend largely upon the rates of progress in different countries.

The ultimate goal of the international co-ordinating bodies (such as ITU and CCITT) and the telecommunications administrations of the member countries is to provide a global telecommunications network employing digital technology and capable of providing a range of transmission services. These include *leased circuit* services and *switched* digital services between individual telephone termination. A network with this switched capability is referred to as an Integrated Services Digital Network (ISDN). The operation of ISDN services over satellite links does present some difficulties, and is being studied on an international basis.

In these respects the United Kingdom is very well placed. The digital conversion programme is at a fairly advanced stage, and digital leased circuit services covering a wide range of speeds from 2.4 Kbits/sec to 2 Mbits/sec are now becoming available in various parts of the country. In addition, pilot trials of ISDN services were scheduled for 1984.

Direct User Access to earth stations is currently used by leased circuit services. However, in the highly competitive US environment, there are at least two examples of specialised satellite carriers that offer to domestic consumers a public switched telephone service by satellite. Such a service is offered by SBS, who utilise the local Bell Company circuits to effect a connection between the customer premises and the nearest earth terminal, for transmission over satellite trunk circuits.

Network Interfacing and Transmission Standards

The CCITT, responsible for drafting standards and securing international agreement, publish recommendations governing such topics as: the interface between user equipment and the network; transmission technology; transmission speeds and classes of service; subscriber numbering; control signalling methods; and other conventions. (Beside transmitting information signals, each end of a transmission link must also exchange *control signals,* to ensure disciplined transmission. A familiar example is the telephone network in which control signals must be generated to initiate such actions as causing the bell to ring and to indicate to the telephone exchange that the call is finished.)

The prior existence of, and continued international support for, transmission standards originally intended to apply to the terrestrial network have contributed significantly to achieving compatibility between the terrestrial terminations of the space link. Nevertheless, there are a number of outstanding sources of incompatibility and other anomalies which have to be resolved in establishing a satellite service and in achieving satisfactory integration with the terrestrial network. The following are some examples:

— the inherent incompatibility of analogue and digital transmission systems. For some considerable time to come, satellite services will have to operate within a terrestrial transmission environment which is part analogue and part digital;

— multiple signalling systems. In many countries several different signalling systems may co-exist in the same national network;

— new modulation techniques. Techniques such as Delta Modulation present difficulties in interconnecting systems in networks employing both analogue and digital links;

— differing digital hierarchies. The existence of a dual standard for the digital transmission speeds hierarchy represented by the North American system using 56 Kbits/sec and 1.5 Mbits/sec (contrasted with the 64 Kbits/sec and 2 Mbits/sec equivalent as employed in the UK and elsewhere);

— differing PCM and speech compression schemes resulting in the possibility that one end of the link might employ 32 Kbits/sec and the other 64 Kbits/sec;

— lack of internationally-agreed standards for video compression codecs, and domestic TV signal encoding and line resolution. It should however be noted that a standard algorithm employing Adaptive Pulse Code Modulation has been developed and agreed in Europe.

Although these problems can be and are being overcome, they nevertheless incur additional engineering costs, and force the adoption of solutions which are expedient rather than ideal.

Service Access Requirements

Two examples are given below which represent different service access requirements. At one extreme, the national earth station exists primarily to interconnect different national services, with traffic comprising bulk telephone channels and broadcast TV. At the other extreme, the business user has modest traffic volumes which can be handled economically by terrestrial access to a shared earth station. Between the extremes are the private earth stations and earth stations supporting specialised services, such as distribution of TV signals to local cable TV networks.

We first describe the infrastructure for routeing telephone calls to the UK national earth station. Attention is then given to the customer access facilities proposed for the European ECS–2 business services. The examples will also be referred to in the discussion of space segment access techniques (Chapter 8).

The UK INTELSAT Gateway

For quite a considerable time the INTELSAT services and access to them were geared to handling bulk telephone traffic, using an analogue speech channel as the basic unit. National earth stations, such as the one at Goonhilly, are linked to the national network at a supergroup or master-group level in the multiplexing hierarchy, and until a few years ago, individual customer access by private circuit was not available. (Figure 7.11 illustrates the arrangements.)

Out-going international calls are collected together via the public telephone network and routed to the international exchange in London. There, a decision is taken on whether calls shall go by satellite or by submarine cable. The traffic destined for satellite transmission is then multiplexed up to supergroup or mastergroup level for transmission to the earth station. From then onwards, telephone channels temporarily lose their individual identity until the channel groups are received and demultiplexed in their destination country. Thus, traffic received in the UK is routed in channel groups to the international exchange, where telephone calls for the UK are routed via the inland network to their destinations. Telephone calls which are in transit via the UK to other countries are separated and re-routed to the appropriate international circuit – generally terrestrial/undersea cable.

In Figure 7.11 there are two earth stations, one pointing to the transatlantic satellite and the other to the Indian Ocean satellite, both of which are accessible from the United Kingdom. In fact, the UK sites at Goonhilly and Madley in Herefordshire are each equipped with several

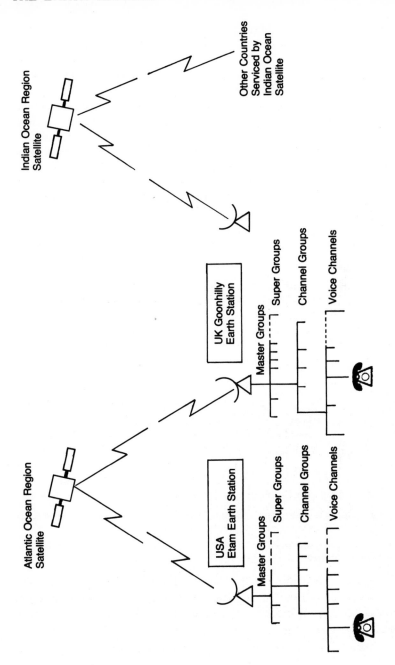

Figure 7.11 INTELSAT Terrestrial Access Arrangements for Trunk Telephony

earth stations accessing a number of satellite systems (including INTEL-SAT and INMARSAT).

The early INTELSAT services employed the FDMA multiple-access technique and although this was cost-effective for bulk traffic, it was not suitable for direct access by users with low traffic requirements. Accordingly, around the mid-1970s, in order to make satellite services economically more attractive for low-density traffic or "thin routes", INTELSAT introduced an additional service called SPADE which employed Single Channel Per Carrier principles (SCPC). This enabled telephone channels to be offered either individually or in low multiples, either on a countrywide basis or to individual users. Prior to the introduction of specialised business services employing small-dish terminals, a number of organisations have been taking advantage of the facility which provides them with a private circuit dedicated to their requirements and directly accessible by leased terrestrial circuits.

Most recently, in response to the possible threat of growing competition on international routes and from regional satellites, INTELSAT has announced its intention of introducing a small-dish specialised business service.

The ECS–2 Business Service

In contrast to the early INTELSAT systems the ECS–2 business service is designed to offer a greater flexibility of access and range of access speeds. Initially, access to the service will be by digital leased circuits to a shared earth station in London. The service employs an SCPC technique, which has been designed to accommodate a range of terrestrial access speeds. The following bit rates are proposed:

— minimum purchasable capacity: 64 Kbits/sec;

— bit rates of $n \times 64$ Kbits/sec (where $n = 1$-30), ie, 64 - 1,920 Kbits/sec;

— 2.048 Mbits/sec.

Although the 2-Megabit rate can be used for data, it is also intended to support video conferencing applications. A more detailed and comprehensive account of business service plans for the UK and Europe is presented in Chapter 11.

8 The Space Segment: Access and Utilisation

INTRODUCTION

This chapter is primarily concerned with the arrangements for accessing the space segment and for sharing its capacity between a number of earth stations. It starts with an overview of earth terminals, with attention to their general features, categories of terminal, and factors affecting their location and performance. This is followed by a description of the earth terminal transmit/receive system. This is in a schematic form which is broadly representative of earth stations in general, irrespective of the specific access methods that are provided.

Attention is also given to Multiple Access and Assignment techniques, including their classification and details of how they operate. The chapter concludes with a review of the main types of control system employed in satellite communications systems.

THE EARTH TERMINAL

General Features

Plates 4 and 5 contrast large receivers from an earth station with a small receiver carried by transporter. Although the difference in scale is striking, appearances can be deceptive. The part which figures most prominently in a photograph or in the landscape is generally the antenna dish and its supporting structure. However, as we shall see there is much more to an earth terminal than this.

In an international gateway station, virtually all the components are contained within the earth station, unlike the situation with shared or private earth stations. In the former, certain functions may be distributed between the two locations; in the latter, all functions will be wholly located on the user's premises. Figure 8.1 is a simplified diagram of a typical customer earth terminal configuration which is marketed as a total package for access to a TDMA satellite service.

The common view that small-dish earth terminals can be readily

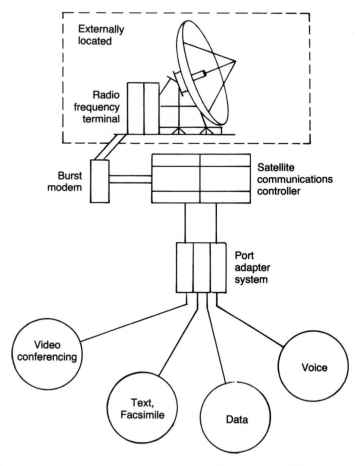

Courtesy of Satellite Business Systems

Figure 8.1   Customer Earth Terminal

located in urban areas (and on the tops of buildings) needs to be qualified. For instance, when mounted on a high building exposed to strong winds, a substantial support frame will be required to minimise movement and possible distortion of the dish system. At ground level, despite the advantages of 14/11 GHz operation, local terrestrial microwave conditions may be such that some form of interference shielding may still be required for the terminal. And finally, judging from the earth terminal siting arrangements in some parts of the USA, some form of concealment may be required to satisfy the environmentalist lobbies or local planning regulations.

Types of Earth Station and Earth Station Standardisation

So far we have distinguished between three types of earth terminal:

— national earth stations or country gateways;

— shared earth stations;

— user or private earth stations.

"Country Gateway" is INTELSAT terminology, and our shared earth station corresponds more or less to INTELSAT's community/urban category. Both community and user earth stations can be further classified according to whether they can transmit and receive, or only receive. A receive-only earth station is much cheaper than one equipped to transmit and receive, and is appropriate for the following kinds of application:

— distribution of TV to cable head ends;

— distribution of TV to town or community antennae;

— transmission of newspapers, magazines, etc for printing at remote sites;

— distribution of information of all types;

— bulk document distribution.

At the present time the only earth station standards which have international force are those defined by the INTELSAT organisation. If an earth station is constructed in conformity with those standards, then it will be capable of accessing the INTELSAT service under the specified conditions.

There are currently five standards – A, B, C, E and F; categories E and F each comprise three substandards – E1, E2, E3, F1, F2, F3. The standards differ primarily with respect to the antenna diameter, the operating frequency, and cost. Types A, B and C cover national gateway requirements, A being the largest (with an antenna diameter of 30 metres), and B the smallest (with a diameter of 11 metres).

The lower cost B and C standards have been introduced for countries whose telecommunications requirements were either limited or at an early stage of evolution. The E and F categories cover the requirements of user, shared and community earth stations. In particular E1 and E2 are designed for accessing the new INTELSAT specialised business services, and have antennae diameters of 3.5 and 5.5 metres respectively.

Earth Terminal Performance

Topics (discussed earlier) such as antennae and link performance, and the various performance figures-of-merit, are directly relevant to earth terminal performance. Because of the very feeble signal that is received, it is important that the antenna and the associated electronics should introduce as little noise as possible. Accordingly, in order to minimise both signal losses and noise in the cables connecting the antenna to the receiver, the first amplification stage is usually built into the antenna; the remainder, if not all, of the radio frequency equipment is located as close as practicable to the antenna assembly. This is shown in Figure 8.1.

The efficiency of the combination is usually quoted as the ratio of the antenna gain to the noise temperature (T), this ratio is an important figure-of-merit for an earth terminal. It is related to the signal-to-noise ratio and is a measure of the sensitivity of the antenna and receiver assembly.

The Earth Terminal Transmission System

Instead of explaining the structure of an earth station in terms of hardware components, it is more instructive to provide a functional description of the transmit/receive path in terms of the features common to all earth terminals. The example given below is broadly applicable to FDMA/SCPC and TDMA systems.

Figure 8.2 depicts the principal functions and corresponding sections of the transmission system.

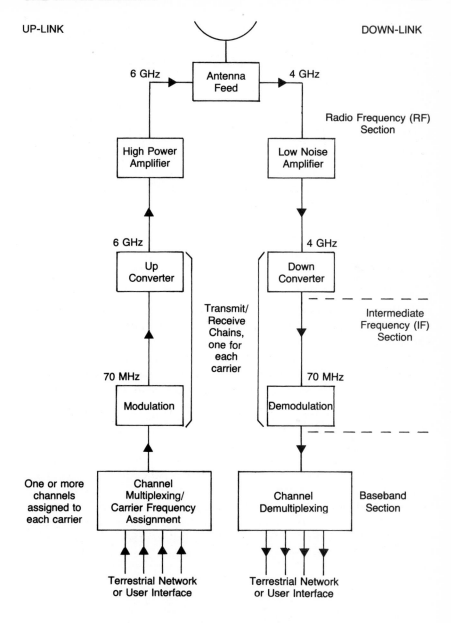

Figure 8.2   Earth Terminal Transmission System

The system is divided into three sections: Baseband, Intermediate Frequency (IF) and Radio Frequency (RF). The functions of each station are briefly outlined below, but first we should consider terminology.

Baseband Terminology

An incoming signal to the earth station is regarded as a baseband signal, relative to the subsequent IF and RF processing stages in the earth station. However, the signal will generally already be in a modulated form so that its characteristics – whether it is analogue or digital, the form of modulation employed, etc – will have been decided within the terrestrial network and/or at the customer's premises.

Channel Terminology

An incoming identifiable channel is likely to comprise a number of subchannels previously modulated and multiplexed together either in the terrestrial network and/or on customer's premises.

In the national gateway context, an input channel comprises a group of telephone subchannels. More generally its significance will be defined by the user; the channel will have a transmission capacity selected from the range of capacities offered by the service. It is also important to observe that transmission in the transmit and receive channels is unidirectional. In order to obtain the equivalent of a telephone circuit or bidirectional circuit on earth, a pair of channels is required across the satellite link. Such a pair is equivalent to a four-wire terrestrial circuit.

Up-Link – Baseband Section

It is not possible to make other than simple generalisations about the particulars of the baseband section. An overriding consideration during the design and implementation of a satellite communication system is to ensure that the satellite system is satisfactorily integrated with the terrestrial network. Apart from this overall constraint, other (general) factors influence the nature of the operations which are performed. These include:

— the type and role of the earth station, eg whether it is a dedicated national gateway station or one offering a specialised business service;

— the terrestrial access arrangements; in particular, whether a multiplexing function is provided at the earth terminal, or whether this is supplied within the terrestrial network or at the customer's premises;

— the space segment access technique, ie whether FDMA, SCPC or TDMA is employed.

Accordingly in the baseband section, the incoming channels may be subjected to multiplexing or demultiplexing operations, resulting in a grouping (or regrouping) of channels into assemblies of multiplexed channels. Alternatively an input channel may retain its original terrestrial identity throughout the transmit/receive chains and across the RF satellite link.

The multiplexing techniques appropriate to the input baseband signal and baseband processing are:

— FDM;

— TDM, including multiplexed packet streams (packet switching terminology is properly regarded as a form of TDM).

Several of the processing operations described in Chapter 7 are performed in the baseband section of the earth terminal, including, in particular, the following:

— Digital Speech Interpolation (DSI);

— encryption;

— Forward Error Correction (FEC);

— voice activation.

The end result of the operations is that one or more channels are assembled into earth station transmit basebands, each assemblage of baseband channels being associated with a radio frequency carrier assigned to the earth station by the *frequency plan*. In SCPC systems each channel will be assigned its own individual carrier, and in multiple-channel FDMA systems, multiple channels will be multiplexed onto a single carrier. On TDMA systems only one carrier/transponder will be present.

Each of the baseband assemblies is allocated its own specific transmit chain prior to modulation onto an RF carrier in the IF section.

Up-Link – Modulation

In the IF section each baseband assembly is then modulated onto an Intermediate Frequency (IF) carrier, and for FDMA systems this is customarily 70 MHz. The reason for using an IF stage is primarily because it is cheaper and more practical to carry out amplification and modulation/demodulation at the lower frequency, whilst the particular choice of 70 MHz is largely historical.

In today's predominantly analogue systems, FDM is the most popular technique for modulating the radio frequency carrier, whilst QPSK is the most efficient for digital systems.

Up-Link – Up Conversion

After passing through a sequence of filtering and amplification stages, the signal is then passed to the up converter which lifts the frequency up to the satellite operating frequency of 6 or 14 GHz (or whatever).

Up-Link – High Power Amplification

The separate transmit chain outputs are then combined together and the consolidated signal is passed to the high power amplifier, and from there fed to the antenna, and onto the up beam. The subsequent transmission route in the space segment is then determined by the transmit/receive chain configuration on-board the satellite.

The Down-Link or Receive Direction

The functions in the down-link or receive direction are largely the reverse of those in the transmit direction. However, because in FDMA systems the earth station also receives carriers with frequencies assigned to other earth stations, it is necessary to eliminate the unwanted carriers. Similarly, in TDMA systems an earth station receives traffic intended for other earth stations, and some mechanism is required to enable an earth station to identify its own traffic.

SPACE SEGMENT ACCESS METHODS

There are two aspects to accessing and utilising a satellite's capacity: the *multiple access* system which enables geographically dispersed locations to access and share the satellite's capacity; and the methods whereby the capacity is allocated or *assigned* to individual users, in terms of both the amount of capacity and the time for which it is made available. We examine the principal *multiple-access* schemes before considering *assignment* methods.

Classification of Multiple-Access Methods

The multiple-access systems used for satellite communications are:

— Frequency Division Multiple Access (FDMA) (Multiple Channel/Carrier);

— Frequency Division Multiple Access (Single Channel/Carrier or SCPC);

— Time Division Multiple Access (TDMA);

— Space Division Multiple Access (SDMA);

— Spread Spectrum or Code Division Multiple Access (CDMA).

The most commonly used technique so far is FDMA together with its derivative SCPC. TDMA is relatively new, but is coming into increasing use. SDMA, whilst its primary purpose is to enable frequency re-use, also serves a multiple-access role, but on its own it does not provide the degree of capacity subdivision generally required, and which can be supplied by the other methods. CDMA we shall not consider in detail, since it is not in common use, although it may have a significant future role in conjunction with access systems which provide random access and capacity assignment for large dispersed end user populations. In practice several different access schemes may be employed on the same satellite.

There is no generally accepted notation for designating the various levels of processing within a multiple-access system, and this is a common cause of confusion. The reader who wishes to delve further into the subject may find the following notation helpful:

A/B/C/D

When read from left to right the symbols specify the sequence of processing in the earth terminal transmit direction from the input channels to the satellite link, and are to be interpreted as follows:

A = The type of encoding/modulation employed for the baseband channel.

B = If the baseband channels are multiplexed before modulation, the type of multiplexing employed.

C = The modulation method employed at the RF stage.

D = The type of multiple access.

We provide some illustrations below.

Frequency Division Access (FDMA)

This was the earliest scheme to be developed and was first employed on the INTELSAT satellites. With FDMA, the bandwidth is divided into smaller bandwidths, and an earth station transmits on one or more of these subdivisions. Within a particular satellite system the subdivisions and associated carrier frequencies are allocated between earth stations and co-ordinated through the Frequency Plan. Allocations can be changed according to demand and changing traffic patterns. Thus, each earth station knows at any time which carriers it will use for transmission and which carriers it will receive on.

Because earth station capacity assignments are fixed (at least in the short term), the routes and route capacities between earth stations are also fixed. The frequency modulation and allied analogue transmission techniques on which it is based had the advantage of being well understood when FDMA was first introduced. Since then it has successfully supported the transmission of bulk traffic on the long haul international routes of the world.

We describe the operation of FDMA, using International Telephony as an example and referring to Figure 8.3 which illustrates an FDMA earth station transmission system.

Up-Link Direction

Voice channels entering from the inland network have previously been grouped into supergroups at the international telephone exchange and sorted according to their transmission routes. These are then multiplexed together, using FDM, onto subcarriers to form FDM baseband assemblies. The individual baseband assemblies are then routed to corresponding transmit chains where they are then modulated onto the 70 MHz IF frequency, using Frequency Modulation.

Up conversion changes each subcarrier into a corresponding preassigned multidesignation RF carrier, and an individual carrier may be

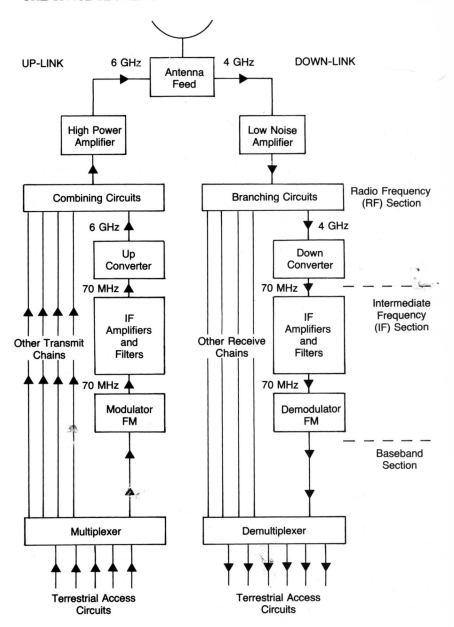

Figure 8.3   Earth Terminal Transmission System for FDMA Access

carrying traffic destined for one or more countries. The different carriers are then combined into a common signal which is fed to the high power amplifier section.

Down-Link Direction

The earth station receives far more traffic than it transmits, and besides the carriers destined for the United Kingdom, it also receives, at the antenna, carriers destined for other countries. Accordingly, following the low-noise amplifier, the unwanted carriers are stripped off, and a branching circuit extracts the required UK carriers and injects them into their corresponding receive chains. From then onwards the transmit procedure is followed in reverse.

In the notation referred to above, this form of multiple access would be referred to as (MM)/FDM/FM/FDMA signifying:

(MM)   = Baseband modulation method.

FDM   = Frequency Division Multiplexing of the baseband.

FM   = Frequency Modulation of the RF carrier.

FDMA   = Frequency Division Multiple Access (of the space segment).

The notation for traditional FDMA is commonly abbreviated to FDM/FM/FDMA.

Figure 8.4 expresses the essential features in diagrammatic form for a system operating three RF carriers. The energy peaks corresponding to the three carriers are depicted in the transponder.

The traditional FDM/FM/FDMA system, as employed by the large INTELSAT earth station, is at its most efficient when using only one carrier/transponder, enabling one transponder to carry about 900 telephone conversations. Such an arrangement is efficient where access is from a limited number of large earth stations-transmitting information large distances over dense traffic routes. However, it is far less suitable for regional communication satellite services accessed by many small earth stations, and for thin low-density traffic routes.

An obvious solution would be to subdivide the transponder capacity into much smaller units of bandwidth, each having its own separate RF carrier. For example, a 36 MHz transponder could be subdivided into 45

KHz segments, which could each carry a single speech channel under FDM/FM/FDMA, and in principle ought to enable one transponder to accommodate about 800 speech channels. However, problems arise as the number of carriers increases. First of all, guard bands are required between each pair of carriers; the more carriers, the more guard bands – with consequent wastage of capacity. Secondly, carriers tend to inter-modulate with each other, and to minimise this effect, the transponder may have to be operated below its full power, so that strong signals do not interfere with weaker ones. Inevitably, the weaker signals win the day. In fact, for a 36 MHz transponder, as the number of carriers increases from

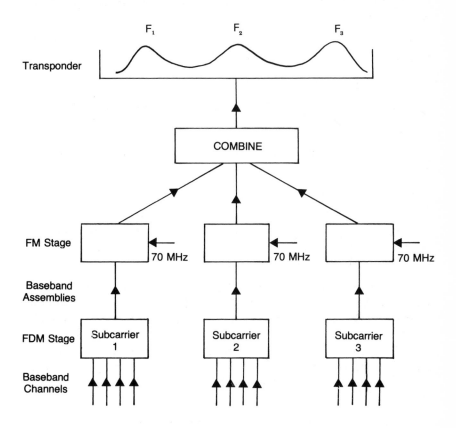

Figure 8.4   FDM/FM/FDMA Transmission Path Showing Three RF Carriers

one to 14, the total number of telephone channels/transponder decreases from approximately 900 to about 330. Loss of power, rather than bandwidth, is the limiting factor on transmission capacity.

A problem facing the designers of satellite communications systems was to find a way to increase the number of earth stations which could share the satellite's capacity without at the same time suffering significant power loss.

Single Channel Per Carrier (SCPC) Systems

SCPC systems are a derivative of FDM/FM/FDMA systems in which the transponder bandwidth is subdivided so that each baseband channel is allocated a separate transponder subdivision and an individual carrier; hence the terminology *single channel per carrier*. In contrast, the FDM/FM/FDMA scheme could be described as a *multiple channel per carrier* system.

Figure 8.5 shows the structure of the transmission path for an SCPC system. In the formal notation the designation would be: (MM)/SCPC/FM/FDMA, where (MM) stands for the modulation method used for the baseband signal. The particular form of SCPC that has just been described employs analogue transmission principles, and is still widely used on thin route satellite communication networks, particularly in Third-World countries. Although analogue SCPC still suffers from the power limitations resulting from the use of multiple carriers, it does enable a larger number of earth stations to access and share the capacity in small, more economic units than can be provided in multiple channel/carrier systems.

However, the efficiency of SCPC operation can be increased considerably by converting to digital transmission, and we shall now consider two examples of digital SCPC operation: SPADE, and the ECS–2 business service.

SPADE

The 'first' operational digital SCPC system was SPADE, which stands for: Single-channel-per-carrier Pcm multiple-Access Demand assignment Equipment. It was designed at COMSAT Laboratories under INTELSAT sponsorship for use on INTELSAT IV and on subsequent INTELSAT satellites. The notably improved performance of SPADE is due to two factors: the use of digital transmission and voice activation. SPADE also has a DAMA capability (discussed later).

By employing PCM at the baseband level and QPSK modulation of the carrier it is possible to accommodate a 64 Kbits/sec voice channel in a bandwidth of 38.4 KHz compared with a full 45 KHz for Frequency Modulation. Using 45 KHz per channel with QPSK, the guard band is effectively included in the 45 KHz, and this enables SPADE to handle 800 voice channels within one 36 MHz transponder.

With FDMA systems it may be necessary to operate at reduced power to minimise interference between adjacent carriers, thereby avoiding a strong signal swamping out a weak one. This wastage of power means that fewer channels can be carried. The SPADE system employs the voice activation technique to conserve power in the following way. When there

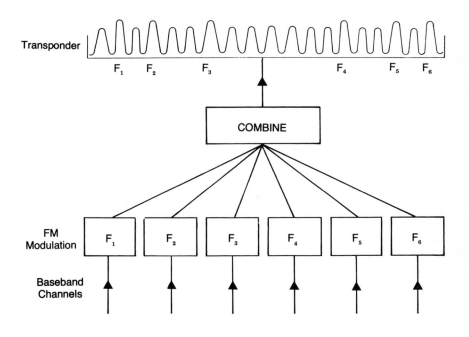

Figure 8.5   SCPC/FM/FDMA Transmission Path

is one voice channel per carrier, the carrier can be switched off when no-one is speaking – a procedure which is impracticable on Multiple Channel/Carrier Systems – and switched on again when speech commences. Bearing in mind that for a two-way circuit, two (one-way) channels are required, this means that even when the full capacity is being utilised, the SPADE carriers can be switched off for half the time, on average.

In our earlier notation SPADE would be represented by: PCM/ SCPC/QPSK/FDMA.

The ECS–2 Business Service

The proposed business service on the ECS–2 satellite will be accessed by digital SCPC using QPSK modulation. A particularly flexible way of subdividing the transponder bandwidth has been defined, and this is illustrated in Figure 8.6.

The total transponder bandwidth of 72 MHz is partitioned into slots which are 27.5 KHz wide, and each slot is able to handle 64 Kbits/sec. A digital customer channel of 64 Kbits/sec or a multiple thereof is allocated respectively either a single slot or the appropriate number of contiguous slots. For an earth segment consisting entirely of 5.5 metre dish earth stations, the business service has a total traffic handling capacity of at least 570 64 Kbits/sec data channels in one transponder. If instead the earth station were of 3.7 metre diameter, the capacity would reduce to 360 channels.

These calculations take into account the reduction in effective capacity resulting from the use of Forward Error Correction, and the required performance standard which is specified using 5.5 metre earth stations.

Each user would be allocated a separate channel unit in the earth station, a fully equipped channel unit providing both a transmit and receive path. The channel units can however be easily configured to support either one-way transmission or reception (such as might be required in a receive-only application).

Time Division Multiple Access (TDMA)

Time Division Multiple Access differs radically from the FDMA schemes considered so far. Instead of a number of earth stations simultaneously sharing transponder capacity, each earth station in turn transmits infor-

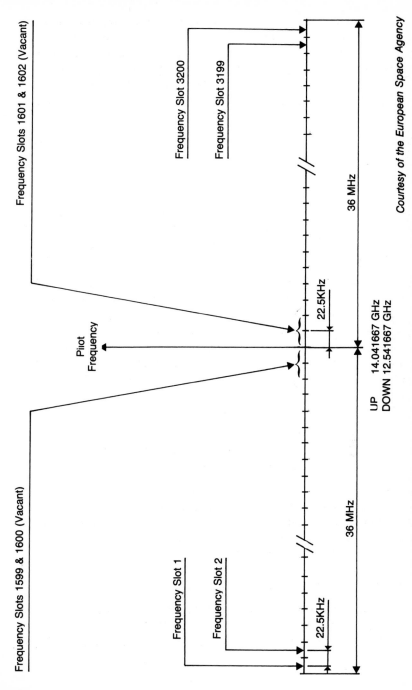

Figure 8.6 Transponder Frequency Subdivision for the ECS-2 Business Service

Courtesy of the European Space Agency

mation in bursts, and has sole use of the entire transponder capacity for a brief duration of time. Figure 8.7 illustrates the process and Figure 8.8 is a schematic of the earth station transmission system. TDMA in effect employs a single carrier per transponder.

Every earth station in a TDMA system receives the entire bit stream and extracts only those bits addressed to it.

TDMA Operating Principles

The operation and control of TDMA systems is considerably more complex than for FDMA systems, and for our purposes it will suffice to outline the main features.

The Frame Structure. The basic repetitive unit in a TDMA system architecture is the *frame* which consists of a number of bursts, one from each earth station connected to the system (see Figure 8.9). In a typical system a frame may have a duration of 15 milliseconds, and using QPSK operating at a burst rate of say 60 Megabits/sec this is equivalent to 900,000 bits/frame.

The first burst in each frame is transmitted by a controlling earth station and this is followed by bursts from individual earth stations, each transmitting in turn. Individual bursts are separated by *guard times* to ensure that adjacent bursts do not overlap or collide at the satellite or at a receiving earth station.

Many frames of similar structure follow each other, and to facilitate the operation of Demand Assignment, in some systems they may also be grouped into master frames, as shown in Figure 8.10.

For the duration of each master frame, an earth station transmits the same burst length and starts its burst at the same time interval after the supervisory burst. At the end of a master frame, a system employing Demand Assignment would then reallocate the earth station burst durations, according to the earth station requirements and traffic activity.

Each of the columns shown in Figure 8.10 represents a continuous stream of bits from one earth station. This stream can itself be split up using Time Division Multiplexing into many separate channels going to different earth stations. Alternatively, individual bursts can be directed to different destinations.

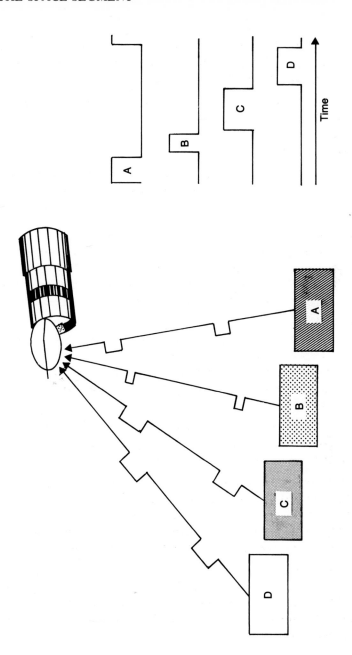

Figure 8.7 Time Division Multiple Access

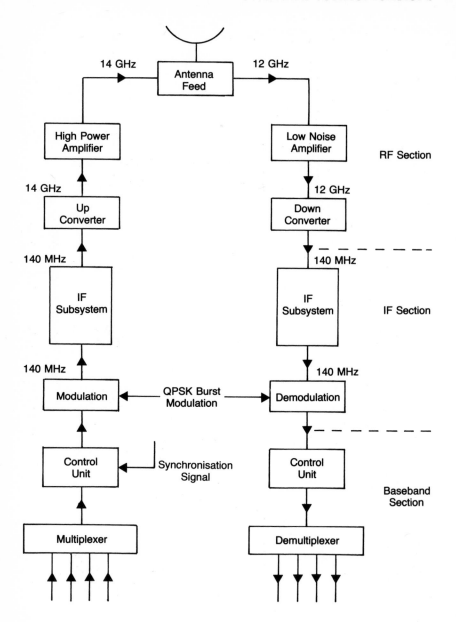

Figure 8.8   Earth Terminal Transmission System for TDMA Access

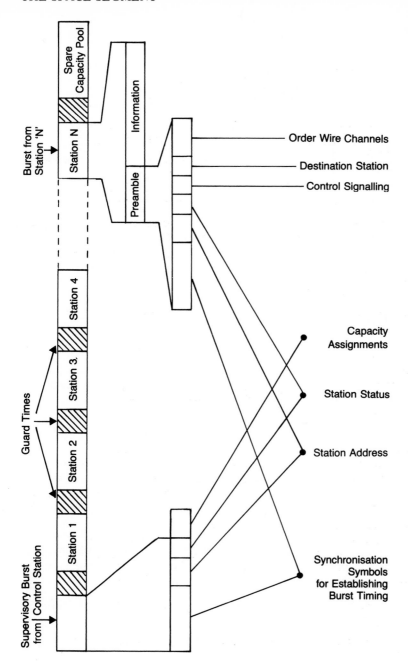

Figure 8.9 A Typical TDMA Frame

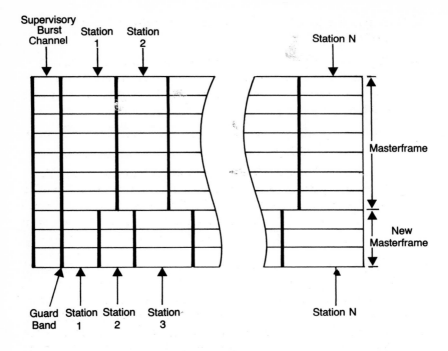

Figure 8.10   Masterframe Showing Time Slot Changed Station Allocations

Burst Structure. In addition to the control information in the supervisory burst, the information field in the succeeding bursts is preceded by a header which also contains control information. The formats of the control fields vary widely from one TDMA system to another, but they serve very similar functions. Referring to Figure 8.9 these are as follows:

— Synchronisation Symbols. To define the start of a frame and synchronise earth stations so that they start their bursts at the correct time referred to the frame start;

— Station Address. To identify the transmitting earth station and the destination earth station;

— · Station Status. To signify whether or not the station is a control station;

— Capacity Assignment. Used by a controlling earth station to respond to a request from an operational earth station for additional burst duration capacity, and to notify of a change in capacity allocation;

— Control Signalling. Used by an operational earth station for various control functions;

— Order Wire Channels. This is an expression originating in North American terrestrial telecommunications practice. In satellite DAMA systems, the Order Wire Channels are used by earth stations for requesting service either in the form of channels in an SCPC system or additional burst capacity in a TDMA system.

Synchronisation. It will be apparent that accurate timing of earth station bursts is crucial to the operation of a TDMA system. Synchronisation is effected by a controlling or reference earth station which emits synchronising signals to all connected earth stations. This establishes a common frame start time for the connected earth stations and enables the successive earth station bursts to be correctly timed relative to the frame start and to each other.

Because of the stringent timing requirements, TDMA systems must employ very accurate oscillators to supply the timing signals. The SBS TDMA system for example uses a rubidium crystal oscillator which enables timing pulses to be generated within a tolerance of ± 1 in 10^{-9} of a second.

The synchronisation mechanism must allow for the fact that the earth stations are located at different distances from the controlling earth station, resulting in variations in the signal propagation time. We also observed in Chapter 4 that for a similar reason TDMA systems impose stringent satellite station keeping requirements.

COMPARISON OF FDMA AND TDMA TECHNIQUES

It is convenient to summarise the respective advantages and disadvantages of the two techniques.

FDMA Techniques

Advantages

Allocation of capacity via the earth station frequency assignments is relatively simple.

No complex timing and synchronisation mechanism is required.

Excepting when the Demand Assignment technique is employed, system management and control are relatively simple.

FDMA earth terminals generally are less costly than their TDMA equivalents.

FDMA is less seriously affected by external jamming and the original signal can be more easily recovered.

Disadvantages

Because multiple carriers/transponder are employed, intermodulation is a problem, and this increases in severity as the number of carriers increases.

To minimise noise resulting from intermodulation, the transponder may have to operate below its full power.

Without using SCPC/SPADE techniques it is not possible to provide Demand Assignment.

FDMA systems lack flexibility in the range of choice of transmission speeds.

The Space Segment transmission paths are pre-assigned, and cannot be readily changed at short notice. This restricts flexibility of use and interconnection potential. In order to have high interconnection capability with other earth stations, an earth station must be able to instantaneously change frequency over a range of frequencies. As we shall see, SPADE with Demand Assignment offers interconnection flexibility, but an earth station needs the frequency agility to support it.

Intermodulation effects and power constraints adversely affect the signal-to-noise ratio and ultimately a transponder's throughput capacity.

Addition of new earth stations to an FDMA system may entail substantial reassignment of frequencies accompanied by administrative inconvenience.

TDMA Techniques

Advantages

The entire resources of the transponder are utilised during each burst, instead of sharing the transponder with other users as in FDMA. Only one signal is ever present at any one instant.

There are no interference problems between adjacent channels caused by carrier intermodulation, since the transponder deals with one RF carrier at a time for a brief interval instead of many as, for example, in analogue SCPC systems. Having only one carrier in the transponder at a time allows the transponder to operate at full power. All transmitting earth stations can operate at full power without saturating the transponder's amplifier, offering a higher signal-to-noise ratio.

The maximum throughput achievable with TDMA is higher than with FDMA, generally by a factor of 2.

TDMA is highly flexible. Channels of widely differing capacities can be intermixed by varying the number of equally spaced time slots allocated to a user. With digital technology, traffic of varying types can also be readily integrated into the same bit stream.

Since each earth station burst carries the address and addresses of the destination earth station(s), TDMA possesses an inherently greater interconnection flexibility than does FDMA.

It is relatively simple to add new earth stations to a TDMA system.

Since TDMA employs digital techniques it also enables Digital Speech Interpolation and other bandwidth conservation techniques to be used.

Disadvantages

TDMA requires very accurate timing and synchronisation.

In order to handle the synchronisation and system control functions, earth stations are more complex and therefore tend to be more costly than FDMA earth stations.

TDMA systems are more susceptible to external jamming and it is more difficult to recover the original signal.

Service Implications

It will be evident that the choice of multiple access systems has an important bearing on the kinds of service than it is technically and economically possible to offer – a view that will be further reinforced when we come to consider Assignment Methods.

At the present time the generally accepted view seems to be that no single Multiple Access technique alone can meet all conceivable requirements. For this reason the majority of communications satellite systems, and particularly those intended to support International, Regional and National Services, employ several different Multiple Access techniques. Optimisation of such a system will need to take account of the following factors:

— required capacity;

— efficient use of the radio frequency spectrum;

— efficient use of satellite radiated power;

— required interconnectivity;

— ability to handle various types of traffic;

— ability to adapt to changes in traffic patterns;

— ease of network reconfiguration;

— terrestrial network interfacing and integration requirements;

— economic viability of alternative approaches.

Table 8.1, based upon one published by CCITT, lists a number of service applications together with the modulation and access techniques which have been used.

ASSIGNMENT METHODS

In a Multiple Access System, users are allocated capacity either in the

Service Application	Baseband Modulation/ Multiplexing	Radio Frequency Modulation	Satellite Access Technique
Telephony Heavy route	FDM TDM	FM PSK	FDMA TDMA
Telephony Thin route	Analogue PCM Delta modulation	FM PSK PSK	FDMA (SCPC) FDMA (SCPC) FDMA (SCPC)
TV Point to multipoint	Video	FM	FDMA
Data High usage, low connectivity network	TDM	PSK	FDMA
Data Thin route	PCM	PSK	FDMA (SCPC)
Data Variable usage, high connectivity network	TDM	PSK	TDMA

*Modification to text

Reproduced from reference 10 in Bibliography by Courtesy of the International Telecommunication Union

Table 8.1 Service Applications and Multiple Access Techniques

form of frequency channels (FDMA) or time slots (TDMA). The Assignment Methods determine how capacity is actively assigned to users in terms of quantity, duration, and whether it is scheduled or varied dynamically, according to demand. The principal ones are:

— pre-assignment fixed: full time, regular part time, irregular scheduled in advance or continuous;

— demand assignment multiple access (DAMA);

— random.

The particular assignment methods which are used have a strong influence on the characteristics of satellite service offerings and how these are presented and packaged.

Although one or other of these methods can be used in conjunction with the Multiple Access Systems we have described, not all permutations are possible. For example FDMA in its multiple channel per carrier form does not readily support DAMA.

Pre-Assignment

With pre-assignment, fixed amounts of capacity are allocated to a user, either on a continuous basis – say 24 hours a day for an extended period – or at scheduled times. The timed basis can be used to accommodate predicted traffic loads at various times of the day or week. Pre-assignment systems in general are best suited for relatively constant traffic. Fixed or continuous assignment systems are particularly associated with bulk long-haul traffic on fixed routes.

Demand Assignment Multiple Access (DAMA)

Whilst pre-assignment methods are efficient and economic for constant and predictable traffic flows, they are generally unsuitable for users or applications with widely-varying traffic volumes and unpredictable traffic flows. Demand Assignment enables the capacities assigned to individual earth stations to be varied dynamically in response to changing earth station demands.

We shall describe two such systems: the SPADE Demand Assignment system, and the TDMA Demand Assignment.

The SPADE Demand Assignment System

Although FDMA, in the form in which it is implemented on International Trunk Services, does not readily lend itself to the application of DAMA principles, it is different with SCPC systems (such as SPADE). SPADE was the first DAMA system to be introduced and possibly the most widely popularised DAMA system.

In SPADE the 36 MHz transponder bandwidth is subdivided into 800 telephone channels, each channel being allocated a separate carrier. Of these, 794 are broken down into 397 pairs to give the equivalent number of two-way telephone circuits. It is important to note that in contrast to multiple channel/carrier FDMA, these individual carriers are *not pre-*

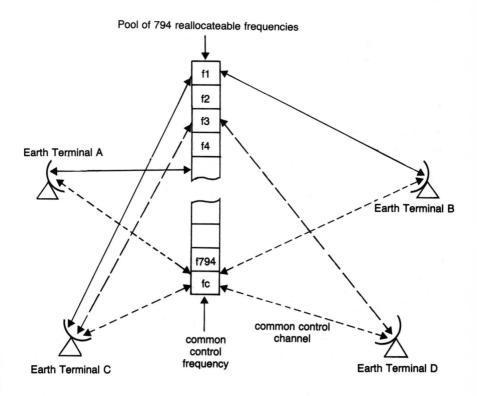

Figure 8.11   Operation of the SPADE Demand Assignment System

assigned to a specific user, and do not have permanent locations associated with their terrestrial terminations. Instead, SPADE treats the 397 pairs as a pool of usable *unassigned* channels. We shall explain briefly how the system operates (see Figure 8.11).

Whenever an earth terminal requires a channel, it seizes a channel from the pool, if there is one free. In the diagram, Earth Station C is using channels f1 and f3 with paths established to earth stations B and D respectively. If C finishes using f3 it will send a signal to all other earth stations to inform them of the fact, and to indicate that this channel has been returned to the pool of available channels.

Station A may then put in a request to use f3 to communicate with station B. Earth Station B would thereupon acknowledge the request, and A and B would then switch their corresponding transmit/receive channels to modems operating at the f3 frequency. In practice a SPADE earth station might be connected to a telephone exchange, with many thousands of incoming lines.

The earth stations communicate with one another through a common signalling channel which transmits at a rate of 128 Kbits/sec. This is obtained by modulating a special carrier at the low end of the transponder bandwidth. This channel has to be shared by all earth stations, and so presents a multiple access problem in miniature, with Time Division Multiplexing being employed. Each earth station is allocated a time slot of 1 millisecond, allowing a burst of 128 bits, and is permitted to transmit a burst every 50 milliseconds. Therefore, one SPADE system can accommodate a maximum of 49 earth stations.

TDMA Demand Assignment

In TDMA systems, reallocation of capacity is effected by varying the slot durations allocated to earth stations. An earth station issues service requests to the reference station via the order wire channel which performs similar functions to the SPADE common signalling channel. The reference station then makes the requested capacity allocation, and notifies the capacity reassignments in the frame supervisory bursts. As Figure 8.10 shows, the procedure takes place in between successive master frames.

In practice, a number of distinct networks may be sharing the same TDMA service, and in order to tailor the allocation of capacity to individual network requirements, each network can be allocated a pool of spare capacity that can be drawn upon by a particular network's constituent earth stations. This is represented as a time slot at the end of a frame in Figure 8.9. Within the individual network context, this can be seen to be equivalent to *transferring capacity* around the network, according to the fluctuating demands from the different earth stations.

Random Access

In this mode of access, instead of the capacity being shared between earth terminals in a tightly disciplined manner, the terminals contend with one another for access, and in effect engage in a free-for-all. They are permitted to send only a short burst of information at a time, but do so at random, and inevitably bursts from different stations will collide and damage each other. However, each earth station can, in principle, detect when this happens and can then arrange to retransmit.

This technique was first investigated and implemented at the University of Hawaii in 1970, and was dubbed the ALOHA system. The name has since achieved an almost generic significance and now refers to a whole class of contention techniques. The ALOHA work also stimulated the extension of these principles to the design of terrestrial Local Area Networks (LANs), of which ETHERNET is a prime example.

ALOHA was developed primarily to support applications in which the traffic arises in sporadic random bursts, characteristic of interactive computing, and it was employed to interconnect a wide variety of terminals and computers via terrestrial and satellite links.

Contention-type systems are not at present in widespread use in satellite communications, because whilst they do work reasonably satisfactorily, they are not terribly efficient, because today's satellites are not really designed to enable contention techniques to be fully exploited. However, they are the subject of intensive on-going research at the present time, since they could provide the appropriate mechanism for enabling thousands of interactive users to access and share satellite capacity. Various developments – such as satellites capable of directing thousands of spot beams to earth, the expansion of cellular radio and cable TV accompanied by the introduction of low-cost earth stations and portable terminals by many observers – are expected to vastly increase the use of satellites for interactive applications.

Miscellaneous Access Techniques

Two further techniques which will be briefly described are "Frequency (or Transponder) Hopping" and satellite switched TDMA (SS/TDMA).

"Frequency Hopping"

In traditional FDMA systems, an earth station is normally assigned a set of frequencies within the bandwidth(s) of one or more transponders, and these remain fixed unless they are changed by agreement. The space segment transmission routes also remain fixed, and are determined by the

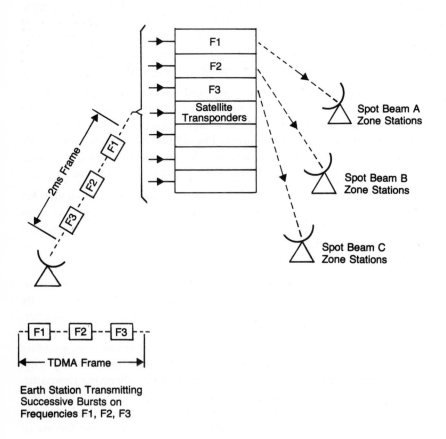

Figure 8.12 "Frequency Hopping"

satellite transmit/receive chains and the antennae beam interconnections.

The operation of a TDMA system over multiple spot beams presents significant problems of co-ordination and synchronisation. These are overcome by "Frequency Hopping" whereby successive burts from a station are transmitted on different frequencies (say F1, F2 and F3) within the TDMA time frame (Figure 8.12). These carriers are received by their corresponding transponders and retransmitted via the spot beam antennae to which the transponders are connected.

Satellite Switched TDMA (SS/TDMA)

Given that an earth station has the necessary frequency agility, frequency hopping enables instantaneous switching of traffic over different space segment routes, and can be employed to give increased interconnection flexibility. An alternative technique which is now under development is to process and switch the TDMA traffic stream on-board the satellite. Initially, switching will be performed using the radio frequency signal but the eventual intention is to provide modulation/demodulation on the satellite and carry out processing and switching using the baseband signal. This will make it possible to restore the signal in space, resulting in an expected two-fold improvement in transmission quality.

The extensive use of SS/TDMA is being proposed for the next generation of satellites planned to come into operation in the latter part of the decade, the European L-SAT series and INTELSAT VI being prime examples.

SYSTEM CONTROL

The overall operation of a satellite communications service involves supervision, monitoring and control functions at three levels:

— monitoring and control of satellite position and performance via the Tracking, Telemetry and Control Link;

— monitoring and supervision of the transmission link and the hardware performance of the earth terminals;

— control and coordination of the Multiple Access and Demand Assignment system.

The terrestrial organisation and the allocation of responsibilities between earth stations vary widely between different systems. The satellite space segment control facility is in general kept separate from the others, although the nominated control earth stations may carry out the functions in addition to their routine operational role. The other two sets of functions may be separated or at the same location. However they are organised, they will intercommunicate with one another and one location may well be identified as the Network Management Centre. Such is the case, for example, with INTELSAT whose global Network Management Centre is located in Washington DC, remote from the satellite controlling earth stations; the same is true of the SBS service.

Multiple Access Systems employing Demand Assignment impose particularly stringent control requirements, since they have to react to changing demands in real time. There are basically two types of control – Centralised or Decentralised – and either can be employed with SCPC or TDMA systems employing DAMA.

Centralised Control

Here the scheduling of capacity is performed at a central location. A computer accepts requests from users, allocates capacity to them, and maintains an up-to-date register of capacity utilisation. The requests and the allocations are transmitted via a common control signalling channel to which all stations listen. The exchange of messages is governed by a special protocol to which all earth stations must conform.

Because the controlling hardware and software are only required at two locations (the principal control earth station, and a standby secondary), centralised control is generally less costly, particularly for TDMA systems. It is likely to be most appropriate for systems which employ elaborate allocation and DAMA schemes.

Decentralised Control

In decentralised control, a registry of channel allocations and channel availabilities must be maintained at all stations. Stations communicate with one another rather than with a central controlling station. A station wishing to transmit to another station selects a free channel and then issues a control message to that station, requesting it to accept transmission on that channel. Again, control messages are transmitted on a common control signalling channel. The intended receiving station then responds, indicating whether it is ready to receive, and if the reply is in the affirmative, the originating station transmits. When the stations have finished with the channel, a control message will be despatched to enable all earth stations to update their lists of available channels. SPADE is an example of a decentralised system.

Because control is decentralised, conflict situations can arise, and these must be catered for in the control protocols. For example, the separate stations may, quite by chance, select the same channel and request permission to transmit. Clearly, rules must be provided to enable this *contention* situation to be resolved. One approach would be to refuse both requests and invite the two stations to re-submit them, possibly separated by a small random time interval.

A major advantage of decentralised (compared with centralised) control is that the system is far less vulnerable to failure. A major disadvantage is its possibly greater cost, particularly if there are many earth stations.

9 Applications Impacts: Echo and Delay Considerations

BACKGROUND

From the viewpoint of the application and its end users, the transmission portion of a communications-based system should ideally be wholly transparent. This means that the system should behave as though the communicating entities – whether people, computers or terminal devices – were entirely localised and not connected over great distances by an intervening transmission medium. In most areas of terrestrial communications, this goal is largely realised. Subject to geographical variations, and the constraints imposed by analogue technology, people can converse over long distances. In well designed data communications systems, to a person using a terminal, the remote computer might just as well be in the same room.

In contrast, satellite communications is a totally different communication environment. In Chapter 2 we identified the unique features of satellite communications: the broadcast property, and echo and delay, the two latter being a direct consequence of the long transmission distances which are involved. These properties have very wide implications for the design of data communications systems employing satellite links. In the interests of efficiency it will be necessary to either modify or abandon techniques and principles appropriate to terrestrial networks. And the broadcast property – which in effect provides total connectivity between all points within the coverage area – is likely to suggest novel approaches to network design. The following are some of the areas that will be affected:

— physical network design implications;

— effect on existing functional network facilities;

237

— efficiency of the existing transmission link protocols, and their enhancement;

— the adequacy of traditional polling disciplines.

Although there is significant accumulated experience of satellite communications in private networks in the USA, a substantial part of the traffic is telephony. And, in data communications, organisations have tended to implement satellite links as a direct replacement for an equivalent terrestrial service. With one or two notable exceptions, satellite service providers have concentrated on supplying the raw transmission channel or transponder capacity, leaving it up to the customer and his equipment supplier to adapt terrestrial based systems to the satellite regime. As a result one encounters installations which are utilising satellite links at less than optimal efficiency.

The emergence of specialised business services together with expansion of demand will encourage the development of packaged solutions to facilitate the efficient integration of satellite links with the terrestrial environment. This needs emphasising because relatively uninformed comment concerning the echo and delay problem has, in some quarters, contributed to the creation of an unfavourable market image for satellite communications. The echo problem for speech has been largely overcome, and the delay problem for data is in principle solvable (or at least can be substantially ameliorated).

In this section we first discuss the effects of delay and echo on speech, and then proceed to consider the implications of delay for data transmission.

SPEECH

Delay Effects

Many readers will have encountered delay when making an international telephone call. Since there is agreement between the PTTs to split traffic between satellite links and terrestrial and undersea cables on a fifty-fifty basis, the odds are that when significant delay was experienced, this was due to the satellite delay.

Transmission delay is inherent in the laws of physics. For speech in particular, there are no practical steps which can be taken to either avoid delay or to minimise its effects. In American organisations with extensive experience of private voice networks using satellite channels, the general impression is that people are aware of it, but accept the limitation and learn to adapt to it.

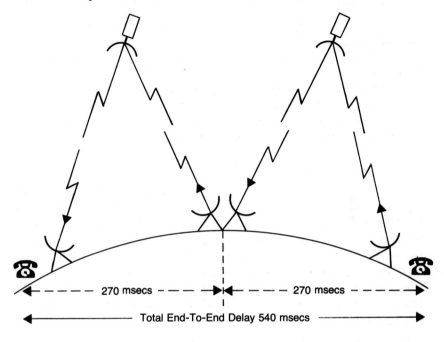

270 msecs

270 msecs

Total End-To-End Delay 540 msecs

Figure 9.1   Single and Double Satellite Hop Delays

A CCITT recommendation places an upper limit of 400 milliseconds one-way delay on a voice connection between any two locations in the world. The one-way end-to-end delay of 270 milliseconds for a satellite link falls comfortably within this limit, provided that only one satellite "hop" is involved. A double "hop" would take 540 milliseconds or thereabouts, thus exceeding the limit (see Figure 9.1). For this reason, international telecommunications services never use routes which entail more than a single satellite hop. For example, incoming transit traffic entering the UK via the transatlantic satellite link, and destined, say, for another country served by the Indian Ocean satellite, would be routed to the international exchange in London, and then re-routed to its destination via a combination of terrestrial and undersea cables. The obvious procedure of merely separating the transit traffic and feeding it to the UK

Indian Ocean satellite earth station is avoided, since this would result in two satellite "hops".

One way of minimising multiple hop delay is to introduce inter-satellite links; this is currently being investigated. Such links would reduce the number of "broken pipes" to earth, and thereby shorten the total transmission path length.

Speech Echo

Echo In Terrestrial Circuits

Here echo is the reflection of a person's voice (ie the speaker hears his own voice transmitted back as he might in a large, empty room).

Although there are many causes of echo in the terrestrial telephone network, the primary cause is the hybrid equipment that is used in any communications system to convert four-wire, long-distance transmission facilities to two-wire, local distribution facilities.

Echo is controlled on circuits shorter than 1,800 miles by introducing loss in proportion to circuit length. This technique decreases perceived echo but still allows callers to understand each other. To control echo when voice signals must travel over long distances (over 1,800 miles) echo suppressors are used. These voice-operated devices constantly determine whether the signals being transmitted are speech signals or echo. Echo suppressors block the transmission of signals that are weaker than the speaker's voice signals, thus acting as a switch which stops the echo, but permits the speaker's words to pass.

Echo suppressors used in the past have generally been analogue devices. With these a delay in blocking echo occurs because of the reaction time of the device in deciding whether the signal is an echo and then activating the switch. This delay may allow some echo to be transmitted before the suppressor blocks it. Conversely, if the switch does not react quickly enough, normal speech signals may be "clipped" or not transmitted.

For example, suppose speaker B interrupts speaker A in the middle of a sentence. Initially, speaker B's speech level may be weaker than speaker A's, and this weak portion of speaker B's conversation may be treated as echo and not transmitted. Consequently, the first word or syllable of the speaker's conversation may be clipped. Although minor clipping does occur, conventional echo suppressors have proved to be a satisfactory solution in controlling echo in terrestrial services.

Echo on Satellite Circuits

Echo is a much more severe problem on satellite circuits because of the long propagation time for a voice signal to reach its destination – approximately 270 milliseconds. However, a further 270 milliseconds have to elapse before the echo is received by the speaker, resulting in a total echo delay time of approximately 540 milliseconds.

The extent to which speakers notice and are affected by echo depends largely on both its loudness, and the time which elapses between the speech and its echo. Because the latter is so much longer on satellite circuits, we can expect it to be far more distracting. The effects of speech "clipping" in "double talk" situations are much more pronounced. Double talk occurs when two people talk at once, or when one speaker interrupts the other. In the absence of effective echo control, speech clipping might occur during double talk situations, since one speaker's voice signals may be treated as echo and subsequently clipped.

The longer propagation delay could lead to a double talk situation when one caller asks the other a question. For example, let us assume speaker A asks speaker B a question. If speaker A does not receive an immediate response, he might assume that he was not heard and might begin to repeat the question just as speaker B's response is being transmitted. Clearly, very few people would relish the permanent prospect of engaging in this form of dialogue.

Echo Control

There are two techniques in use for controlling echo on satellite circuits: echo suppression and echo cancellation.

Echo Suppression. As we saw when discussing echo control in the terrestrial context, echo suppression suffers from an inherent conflict between removing echo on the receive path and yet allowing bona fide speech to be received, without clipping.

Echo suppressors designed for use on satellite channels are a considerable improvement on those traditionally employed in the terrestrial network. They employ digital technology, and apart from having several other advantages, they have a far faster reaction time than analogue suppressors. At least one model of suppressor is available which can be programmed to allow a weaker speech signal to break in sooner, thereby reducing the amount of speech clipping.

Echo Cancellation. Echo cancellers overcome the inherent conflict present in echo suppressors by automatically synthesising a replica of the expected echo on the receive path, and then subtracting it from the echo when it is received, thus cancelling the echo component of the received signal. Neither direction of transmission is interrupted and double talking can normally occur without clipping. This procedure is illustrated in Figure 9.2. Of the two techniques, echo cancellation is the more efficient and is increasingly favoured.

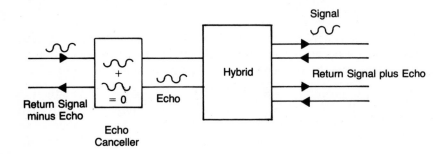

Figure 9.2   Echo Cancellation

Echo suppressors interrupt the communications channel, and it is necessary to temporarily disable them when a voice channel is needed for data transmission. The desirability of disabling echo cancellers for data transmission is currently under study by CCITT.

In conclusion it is worth observing that if the communicating telephones each had direct access to the satellite link, the echo problem would disappear, since they would then be connected by a four-wire circuit. End-to-end four-wire transmission will eventually be realised with the introduction of the Integrated Digital Services Network (ISDN), in which the existing two-wire local distribution circuits will be replaced by four-wire circuits. However, and until this is achieved, some form of echo control will be required.

DATA TRANSMISSION

In common with its terrestrial counterparts, a satellite link transmits information in the form of electrical impulses, and has no knowledge of either the significance of the information transmitted or the application to which it relates. From the point of view of the end-user applications, it is essentially a passive vehicle, although possessing certain attributes and limitations. In order therefore to initiate and sustain meaningful inter-communication, it is necessary to superimpose a set of externally defined rules or *protocols*. The protocols not only enable the link to be utilised, but they also, in effect, mask the underlying transmission characteristics from the application. In consequence, the protocols which are employed have a strong influence on the performance of the link as seen by the communicating entities and end users. Depending upon the design of the protocol, it can enhance performance and compensate for any deficiencies in the physical transmission vehicle.

A protocol, necessary (for example) for making a telephone call, includes such items as: call initiation and dialling procedure; specification of the ringing signal; and speaking and linguistic conventions. Unlike telephony, however, the protocols required for data transmission are far more complex. And, because of their performance implications, protocol considerations are unavoidable in any meaningful discussions of the effects of link delay.

Protocol Functions

General

In any communications system, several levels of protocol are necessary. These are generally arranged with user application protocols at the highest level and data link control protocols at the connect level. Here we are primarily concerned with the data link control level. The protocols themselves are implemented by combinations of hardware and software at each user location.

A protocol can be regarded as a specialised and self-contained communications system in miniature. It comprises a set of rigorously defined procedures which will inititate and perform specified actions, either on request, or in response to some event. The operations are controlled by means of messages which are exchanged between the remote ends of the link. (Readers wanting more information about protocols are referred to references 1, 5 and 11 in the Bibliography.)

Much of the investigational and experimental work on protocols for satellite communications, particularly for high volumes and high bit rates, has been conducted by the staff of SBS and COMSAT Laboratories. For further information, the reader is directed to references 12 and 16 in the Bibliography.

The most important functions and properties of a data link control protocol are discussed below.

Link Initiation and Termination

Until entities wish to communicate, the path joining them is essentially in a quiescent but (usually) well-defined state. By some appropriate signalling convention, one or other part must be able to 'seize' the link and initiate the interaction. Similarly, there must also be prescribed rules for terminating the interaction.

Synchronisation

Since the communicating entities are essentially autonomous and operate according to their own particular rules, it is clearly necessary to synchronise the temporal sequence of events. Synchronisation will normally be required at several levels. Because information is transmitted serially in bits, bit synchronisation will always be necessary. Moreover, since information is generally given a character structure superimposed on the bit stream, synchronisation of character boundaries will also be necessary. Depending upon the form of transmission and protocols employed, the characters may also be grouped into larger units, such as blocks or packets (in packet switching networks). Almost always, the application program will only be interested in the actual message, or some other structure, such as a file, meaningful to the application.

Link Control

In relation to both the initiation of the interaction and its subsequent progress, the question arises of the relationship between the two distant ends. Is the relationship a subordinate one, one party being in full control and the other essentially passive, or can they function as equals?

The polling mechanism employed in multidrop configurations is an example of the first approach, and is a dominant feature of traditional computer communication networks. Here, to avoid contention between a number of devices sharing the same circuit, each device must receive an invitation before it is permitted to transmit. In a different version, on packet switched networks, the network in effect continuously polls the connected devices, which can then transmit without being invited to do so, although the network will reject the transmission if it becomes congested, or for other reasons.

In point-to-point links, the relationship tends to be one of equal parties, and is becoming increasingly important in distributed processing networks.

Error Detection and Correction

Transmission technology continues to improve and the introduction of digital transmission is resulting in dramatic improvements in performance. However, the transmission network can never be completely error-free. Therefore a substantial part of the protocol is likely to be concerned with the methods of recovery from errors, whether caused by faulty operation in one of the interacting devices or faulty operation in the system which connects them. Much of the increased complexity of the more recently developed protocols is due to the improved error-detection and error-recovery mechanisms. An important requirement is to place responsibility for error detection and recovery as far away as possible from the application levels of the system. This has the advantage that the application is largely insulated from errors, and most of the time it will be entitled to assume that most of the information that it receives is error-free.

However, a price has to be paid for enhanced error control. This involves the small but finite probability that blocks of information get lost, are duplicated, or fail to arrive at their destination in the same order in which they were despatched by the sender.

Flow Control

The main objectives of flow control are to ensure that:

— blocks of information arrive at their destination in the same sequence in which they originated;

— information will be accepted and delivered at rates which match the capacities of devices/users at the transmitting and receiving locations respectively;

— subject to the above requirements, the utilisation or throughput of the channel will be maximised (ie information is transmitted at a rate as close as possible to the rated speed of the channel).

A question of great concern is that the progress of an interaction may come to an unintentional halt. This may occur because each party, following the rules strictly, is waiting for a message from the other one or, for some other reason, such as storage limitations, is unable to respond. Various mechanisms are employed for handling these situations. The imposition of a *time out* is one way, so that if one end does not receive an acknowledgement within a specified time interval, it aborts its previous transmission, thus avoiding a deadlock situation. (Here each party is waiting for a response from the other party without which the interaction cannot proceed.)

Protocol Resilience

The modern protocols now coming into use are very sophisticated and cater for a very wide range of possible circumstances. Nevertheless it is currently impossible to fully test them over what is, for practical purposes, an infinite range of possibilities. Accordingly, these protocols make provision for one or the other party to effect a restart in a situation which appears to be totally deadlocked or impossible to recover from.

Efficiency and Cost Effectiveness

Ideally a protocol must make efficient use of the channel capacity, and should aim to maximise throughput whilst at the same time delivering information which, so far as possible, is error free. And, of course, it should achieve this at an acceptable cost. The implementation of modern data link protocols not only requires substantial computing capability at the remote locations, but significant and possibly large amounts of buffer storage will also be required to support the error correction and flow control procedures.

Performance and the Satellite Transmission Environment

Before discussing the implications of satellite delay on applications and system performance, it is useful to list performance criteria against which system and protocol performance can be judged, and to identify the characteristics of the satellite transmission environment which influence performance. These are presented below.

The Transmission Environment

Traffic pattern: interactive/bursty; bulky/continuous. The distinction between traffic categories influences the choice of performance criteria. For interactive traffic, *response time* is particularly important, and for bulk traffic, *throughput efficiency.* Ideally, a protocol ought to be able to handle both with equal facility, although its design may not be an easy task.

Volume of data bits/transmission. Of importance in accessing throughput and error detection performance at high volumes/high bit rates. In their investigation, SBS considered transmissions each of 10^{10} bits.

Transmission Rate. Theoretical and experimental investigations to date have covered all speeds up to 6 Mbits/sec, although attention has been mainly focused on the range 56 Kbits/sec to 6 Mbits/sec.

Transmission Block Size. This has a major influence on protocol throughput efficiency. For a given set of transmission and performance parameters there is generally an optimum block size.

Link Delay. Almost all the investigations (and also the conclusions and recommendations of this chapter) apply to single-hop transmission paths. The effects of double- and multiple-hop delays are currently under investigation.

Transmission Channel Bit Error Rate (BER). Refer to Chapter 6 for the definition and comments on channel Bit Error Rates.

Performance Criteria

Response Time. This is defined as the elapsed time from depression of the user terminal transmit key until the first character of the return message is received. This is a particularly important parameter in the design of interactive systems.

What is considered to be a satisfactory response time is decided partly by the psychological needs of the user, and the application. The trend towards more powerful display units, graphics terminals, etc and more time critical applications is imposing more demanding requirements. An average response time not exceeding two seconds is becoming increasingly common in interactive applications.

Throughput Efficiency, defined as:

$$\frac{\text{Actual Transmission Rate} - \text{Achieved (bits/sec)}}{\text{Nominal Speed of the Channel (bits/sec)}}$$

In the investigations referred to earlier, SBS were seeking protocol improvements which would offer throughput efficiencies of 90% or greater, for 10^{10} bits transmission volumes.

Residual Error Rate. This refers to the proportion of errors which are transmitted and received undetected. In relation to the transmission file sizes referred to above, SBS used the criterion that no more than one undetected error should be received in 100 transmissions (each of 10^{10} bits).

Throughput Efficiency/Transmission Speed Independence. This is a very desirable property, if it can be achieved, since the protocol is operating optimally so far as transmission speed is concerned.

Minimum Sensitivity to the Channel Bit Error Rate (BER). Because the BER is subject to unpredictable fluctuations, and can also vary between different earth station links, it is desirable that a protocol should so far as practicable be relatively insensitive to changes in the BER.

Satellite Delay Implications

The data link protocols currently in use in terrestrial communications are the result of many years of development, and form the basis of a number of ISO (International Standards Organisation) Standards or CCITT Recommendations. Whilst most of these work satisfactorily in terrestrial communications – and the modern ones such as HDLC (High-Level Data Link Control) are particularly efficient – they are likely to perform far less satisfactorily when implemented in an unmodified form over a satellite link. The degree of performance impairment partly depends upon the particular class of application. In the case of some interactive terminal applications, the performance might be tolerable. However, in other cases – in particular large volume data/file transfer at high bit rates – a severe degradation in performance may be expected.

We now examine the main implications of delay for data transmission, and for data transmission protocol design.

Response Time

The delay for a one-way transmission between earth stations approximates to 270 milliseconds or 320 milliseconds if a 50 milliseconds allowance is made for processing. Accordingly the response time from depressing a terminal transmit key to receiving the first character of the reply approximates to the round trip delay of 640 milliseconds, and this clearly makes a significant inroad on a response time of the order of two seconds.

Throughput

If a satellite link is substituted for a terrestrial circuit without any changes to the system's protocols, software and hardware, there can be a significant degradation in throughput efficiency – *reducing perhaps to a half or less of that experienced on the equivalent terrestrial circuit.* The effect becomes increasingly pronounced at high transmission speeds.

Polled Terminals

Polling can give efficient usage of terrestrial channels, and is probably the most widely used line control discipline for interactive terminals. Polling on terrestrial networks is generally most efficient when the polled line has a fast turnround, but becomes less efficient as the number of polled locations and propagation times increase. With a round trip delay of 640 milliseconds, polling of remote terminals via a satellite link must be regarded as inappropriate.

Several alternatives to remote polling via satellite have been considered. One approach is to operate the terminals in a purely local mode, but this would only be suitable for terminals close to an earth station. Another approach is to poll remote terminals from a terrestrially located concentrator which then interfaces with an earth station. In either case a different protocol would be employed across the satellite link. It is claimed by a number of authorities that the latter procedure provides an acceptable alternative.

Flow Control

An important function of flow control is to match the speed of the transmitting device to that of the receiving device. At the individual data link level, flow control is the responsibility of the data link protocol, but for a computer network in its entirety, overall flow control will come under the supervision of the particular network functional architecture which is employed. Examples of proprietary functional architectures include: Digital Equipment Corporation's DECNET, and IBM's SNA.

Flow control regulates the flow of traffic so that both ends of the link keep in step, and any disparity which arises is accommodated by buffers at each end of the link.

The flow control parameters (including the buffer sizes) are determined by such factors as: the link propagation delay, the range of transmission speeds to be employed, and the speeds of the communicating devices. The great majority of communication-based systems employ data link protocols and network architectures whose parameters are adjusted to the terrestrial transmission environment. However, with the satellite delay inserted into the link, these parameter settings can be expected to play havoc with performance.

Time-Outs

For protocol implementation on terrestrial circuits the time-out threshold values will in general be appropriate to the propagation delays encountered on terrestrial networks. These are unlikely to operate efficiently over satellite links, particularly if simple ARQ protocols, which permit only one frame or block of information to be outstanding at any one time, are employed. A great deal of valuable channel time may be wasted, and in extreme cases the lengthy delay can cause the protocol software to go into a loop condition, resulting in an irrecoverable deadlock situation.

Error Detection and Correction Procedures

The particular form of error detection and correction procedure employed has a profound impact on protocol efficiency, not only with regard to the residual undetected error rate, but also with regard to throughput performance. It arises in this way. A detected error, and its correction, creates an interruption in the flow of data, and the two issues of crucial significance are: firstly, how long does the interruption last, and how is the transmission time utilised? And secondly, what happens to the blocks which may be transmitted immediately following the detection of the block in error?

In block-by-block or Idle-ARQ protocols, transmission ceases until the block has been retransmitted and correctly received. The elapsed time that this takes is equivalent to the round trip delay of 640 milliseconds, during which a substantial amount of data could have been transmitted. As Table 9.1 shows, this amounts to about 17 Kbits and 2 Mbits at transmission speeds of approximately 56 Kbits/sec and 6 Mbits/sec

Transmission Speed	Bits in Transit	
	One Way	Round Trip (Duplex Transmission)
56 Kbits/sec	17.92 Kbits	35.84 Kbits
112 Kbits/sec	35.84 Kbits	71.68 Kbits
224 Kbits/sec	71.68 Kbits	143.36 Kbits
448 Kbits/sec	143.36 Kbits	286.72 Kbits
1.544 Mbits/sec	494.08 Kbits	988.16 Kbits
3.152 Mbits/sec	1.01 Mbits	2.02 Mbits
6.312 Mbits/sec	2.02 Mbits	4.04 Mbits

Table 9.1   Data Bits in Transit Over a Satellite Channel

respectively. Similar arguments apply to protocols which permit continuous transmission but discard blocks received following an error, even though they may in fact be correct. There is some resulting loss of throughput. Accordingly one central objective of protocol design over the years has been the development of more efficient error control procedures.

Protocol Alternatives

We shall now review the principal types of data link protocol and their suitability for satellite transmission.

Information Feedback

Sometimes referred to as "echoplexing" this is the most rudimentary technique employed on unintelligent slow-speed teletype compatible terminals. As each character is entered on the keyboard it does not directly activate the local printer but is transmitted from the keyboard to the receiving end, which loops the signal back on the return path of the duplex pathway. When the returned signal reaches the originator the character is displayed on the local printer. Error detection consists of the operator checking to see whether the printed character agrees with the one which was input. With a round trip delay of approximately 640 milliseconds for each returned character, this is not surprisingly regarded as unacceptable, as shown in the following example:

Terminal speed: 30 characters/second

Calculations on a per character basis

	Satellite Delay (secs)	Transmission Time/Character (secs)	Total Time/Character (secs)
Input	0.320	0.033	0.353
Echo	0.320	0.033	0.353
Total Time/Character			0.706 seconds

Terrestrial for comparison = 0.20 – 0.50 seconds
Conclusion: Unacceptable

The preceding is commonly referred to as duplex operation for this class of terminal but it can also be operated in a half-duplex mode by suppressing the echo back facility. Here the results are somewhat more encouraging:

Terminal Speed: 30 characters/second
Input Message Length: 20 characters
Received Message Length: 40 characters

	Message Transmission Time (secs)		Satellite Delay	Total Service Time
	Input Message	Return Message		
Terrestrial	0.67	1.33	–	2.00 + computer time
Satellite	0.67	1.33	0.64	2.64 + computer time

Conclusion: Satellite response is 0.64 seconds slower

Binary Synchronous Protocols

The first generation of protocols designed for synchronous transmission channels are referred to collectively as binary synchronous or block-by-block protocols, and employ the ARQ (Automatic Retransmission on Request) technique. The IBM Bisynchronous Link Control and the ISO Basic mode are representative of this class of protocols. The ARQ technique employed in these protocols operates in the following way.

The transmitter is allowed to send one block of information, and no more, and must wait for an acknowledgement (ACK) from the receiver to confirm whether it has been received correctly or not. The receiver establishes the correctness by recalculating a block checksum and comparing it with a transmitted checksum.

If the block has been received correctly then the transmitter can send a new block. If not – signified by the receipt of a NAK – the transmitter must send a fresh copy of the block. The procedure is illustrated in Figure 9.3(a). The shaded areas in the diagrams in Figure 9.3 represent idle transmission time, and in Figure 9.3(a) this is due to the transmitter waiting for an ACK and the receiver waiting for a transmission.

Idle time results in a corresponding loss of throughput, and, as transmission speeds increase, throughput efficiency progressively diminishes.

There are various implications for satellite transmission performance:

— for small block sizes, the transmission efficiency is low, because the block transmission time is short compared to the block transmission cycle, reflecting the loss of efficiency due to satellite delay;

— for very large block sizes, the efficiency is low because of the frequent need to reject large blocks with one or more errors in them;

— for the data values which have been investigated, transmission efficiencies of 90% or better with BERs as high as 1 in 10^5 cannot be achieved with this class of protocol. Indeed, it can be shown that a BER of 1 in 10^9 or lower is needed to achieve this level for the transmission rates of interest. Transmission efficiency is also very sensitive to fluctuations in the BER.

Continuous ARQ Protocols

With the objective of improving the efficiency of terrestrial link pro-

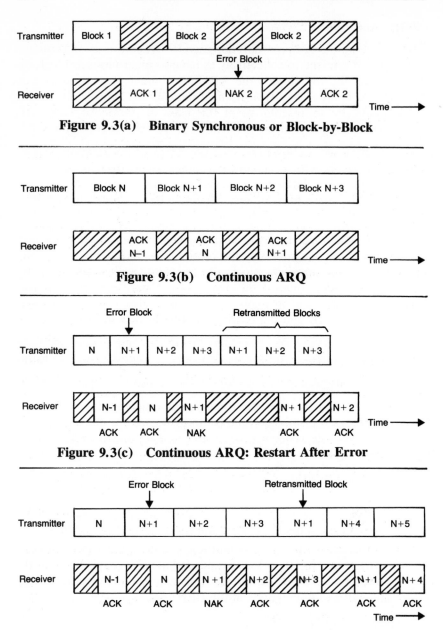

Figure 9.3(a) Binary Synchronous or Block-by-Block

Figure 9.3(b) Continuous ARQ

Figure 9.3(c) Continuous ARQ: Restart After Error

Figure 9.3(d) Continuous ARQ: Selective Retransmission

tocols, a new family of protocols (of which HDLC, SDLC and ADCCP are representative) was developed. These employ a technique called continuous ARQ – shown in Figures 9.3(b), (c), (d) – which operates in the following way.

A number of data blocks are transmitted in succession by the transmitter without waiting for individual block acknowledgements. At the receiver they are processed and acknowledged as in the block-by-block procedure. However, by the time the transmitter gets an acknowledgement, it will generally have transmitted further data blocks, so that acknowledgements will lag behind the transmitted data blocks.

Unlike block-by-block a number of blocks can be awaiting acknowledgement, and multiple blocks can be acknowledged by one acknowledgement message. The precise number of blocks which can be outstanding at any instant is specified as a parameter and is called the "window". For terrestrial applications, a value of 7 is commonly chosen, and in anticipation of the increased number of blocks which could be in flight simultaneously on satellite links, and therefore awaiting acknowledgement, ISO, the sponsoring organisation, introduced an extension to the standard to allow a window size of 127. For error detection this class of protocols employs the powerful cyclic redundancy checking mechanism.

In order for the procedure to operate at all, some additional facilities must be present. First of all, at the transmitter each block transmitted but not acknowledged must be stored in case retransmission is required. Secondly, at the receiver, if a block is found to be in error, then a number of subsequent blocks may be received before the block in error can be retransmitted, and this may disturb the sequence of the data blocks. Therefore in order to control sequence and maintain a correct correspondence between blocks and their respective acknowledgements, it is necessary to include in each block a unique block sequence number.

In continuous ARQ there are two different techniques for handling retransmission: Re-start After Error and Selective Retransmission.

In *Re-start After Error,* when the receiver sends a NAK, indicating that the block with the corresponding sequence number must be retransmitted, the sender transmits that block, and then continues transmitting subsequent blocks, even though they have already been transmitted. The procedure is shown in Figure 9.3 (c).

Following receipt of the erroneous $N+1$ block, the receiver discards blocks $N+2$ and $N+3$ since they are out of sequence, and waits for the retransmitted block $N+1$ to arrive. When block $N+1$ has been transmitted, the transmitter continues with blocks $N+2$, $N+3$, and so on. Each end is able to keep track of the situation by reference to the block sequence numbers.

This is the form in which HDLC and related protocols are currently being implemented on terrestrial networks, and with a window size of 7, they offer a superior performance compared with block-by-block protocols.

Whilst this type of protocol automatically preserves the order of data transmitted it does so at the expense of discarding the data in transit between a detected error and when the transmitting station receives the NAK. Therefore there is a loss of the transmission which occurs during the round trip delay. At a transmission rate of 6.132 Mbits/sec this amounts to about 4 Mbits of data which is lost each time a NAK is issued by the receiver. Table 9.1 gives the losses at lower transmission rates.

Duplex transmission must be employed, otherwise the protocol degenerates into a block-by-block type.

Transmission efficiencies of 90% or better with channel bit error rates as high as 1 in 10^5 cannot be achieved.

Transmission efficiency is a function of the transmission rate. It is also very sensitive to fluctuations in the BER for rates as high as 1 in 10^5, particularly for transmission rates above 1.544 Mbits/sec.

Since the probability is high that most of the discarded data could have been received correctly, it follows that this protocol has a built-in inefficiency, particularly at high transmission speeds.

The distinction between low and high transmission speeds is somewhat arbitrary but US satellite operators tend to treat any speed below 56 Kbits/sec as low and any other speed as high. However, since continuous protocols like HDLC are still considerably more efficient than the block-by-block class, it is worthwhile considering how HDLC would perform over a satellite link at a typically "low" terrestrial speed such as 9,600 bits/sec. Figure 9.4 illustrates its performance alongside a binary synchronous protocol for comparison.

This shows that for a binary synchronous protocol, the throughput is greatly reduced on the satellite link, and with a frame size of about 1000 bits, performance reduces to one third. The diagram for HDLC, however, shows that this gives far better performance on the terrestrial circuits. Furthermore, if a satellite link is substituted, there is a negligible drop in throughput providing that a frame size greater than about 1200 bits is employed. There is a slight fall-off in performance as the BER increases from 1 in 10^7 to 1 in 10^5. This is achieved with a window size of 7, and by increasing this to 127, a similar result can be obtained with a much smaller frame size.

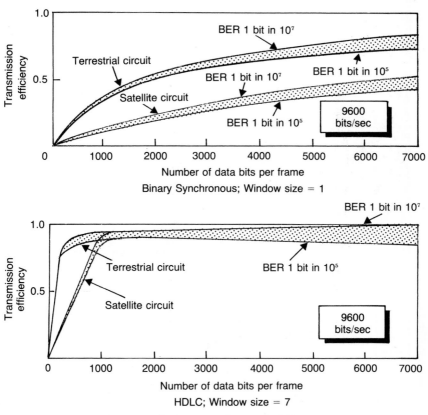

James Martin, COMMUNICATIONS SATELLITE SYSTEMS, © 1978, p 288
Adapted by permission of Prentice-Hall Inc, Englewood Cliffs, N J

Figure 9.4   Comparative Performance of Binary Synchronous Protocol and HDLC on Terrestrial and Satellite Circuits

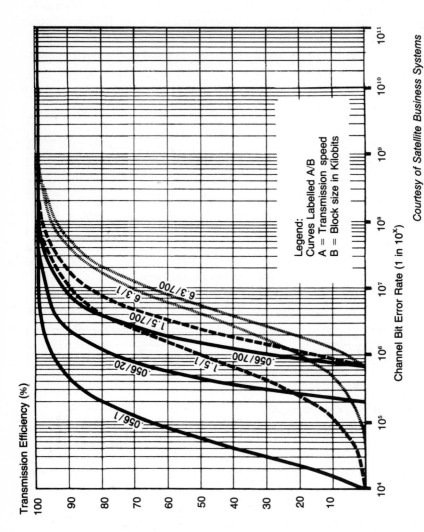

Courtesy of Satellite Business Systems

Figure 9.5   Restart After Error: Transmission Efficiency vs Channel Bit Error Rate (BER)

In order to illustrate how the protocol behaves at higher transmission speeds we present some results of the SBS investigations. Figure 9.5 plots efficiency against BER for various combinations of transmission rate and block size. This shows for example that, for a BER of 1 in 10^5, a transmission speed of 56 Kbits/sec and a block size of 1,000 bits, an efficiency of about 65% is achieved, and at a BER of 1 in 10^6, 95% is achievable. At higher transmission rates the efficiency falls off dramatically, unless the channel BER is substantially improved. In fact, as Table 9.2 demonstrates, in order to achieve transmission efficiencies better than 90% over all transmission rates of interest, a BER lower than 1 in 10^7 is needed.

Several points should be noted in interpreting these results:

— the performance criteria referred to earlier have been applied, ie file transmission comprising 10^{10} bits and an undetected error probability of 0.01;

— although the basic HDLC frame structure and principles of operation have been retained, certain modifications have been introduced;

— in particular, it has to cater for window sizes far in excess of the recommended 127. This is to allow for the greatly increased number of frames that can be in transit, and awaiting acknowledgement.

Transmission Rate (Mbits/sec)	Maximum Achievable Efficiency For Error Rates 1 in 10^X						
	x = 4	x = 5	x = 6	x = 7	x = 8	x = 9	x = 10
0.056	19.09	72.50	96.53	99.63	99.96	99.99	99.99
0.112	11.48	57.79	93.22	99.28	99.93	99.99	99.99
0.224	6.11	40.77	87.38	98.49	99.86	99.98	99.99
0.448	3.16	25.66	77.64	97.21	99.71	99.97	99.99
1.544	0.94	9.12	50.25	91.00	99.02	99.90	99.99
3.152	0.47	4.72	33.30	83.33	98.04	99.80	99.98
6.312	.23	2.40	19.82	71.22	96.12	99.60	99.96

Courtesy of Satellite Business Systems

Table 9.2   Transmission Efficiency vs Channel Bit Error Rate (Block size: 1,000 bits)

In general, window size is a function of the block size and the transmission speed. For example, at a transmission speed of 56 Kbits/sec and a block size of 1,000 bits, about 18 blocks can be in transit, and at 6 Mbits/sec, 2,000 blocks can be in transit. However, for a given set of transmission and performance parameters, there is in general an optimum block size.

A fundamental deficiency of Restart After Error is that a large number of blocks may need to be retransmitted, thus wasting channel capacity. By contrast, in *Selective Retransmission,* illustrated in Figure 9.3 (d), the transmitter only retransmits block N+1, the block in error, whilst the receiver accepts blocks N+2 and N+3, even though N+1 has to be retransmitted and will therefore be out of sequence when it is received. After retransmitting block N+1, the transmitter continues with blocks N+4, N+5, etc.

As each block is received by the receiving terminal, it is checked for correctness and acknowledged. If no error is detected in the received block, a positive acknowledgement (ACK) is sent to the transmitting terminal and the block is forwarded to the receiver's attached equipment. If an error is detected in the received block, the block is discarded and a negative acknowledgement (NAK) is sent to the transmitting terminal. However, for this protocol, the receiving terminal continues to receive, check, and store correctly-received blocks until the (NAK'ed) block is retransmitted and received correctly. Then the receiving terminal releases the received data (in order) to the destination device.

When the transmitting terminal receives a NAK for a particular block, it continues transmission, locates the block received in error and retransmits it as part of the block stream. This continues until all data is received correctly at the receiving terminal and positively acknowledged. However, the protocol does entail restoration of the block sequence at the receiving terminal.

There are various implications for satellite transmission performance:

— with this protocol, all data correctly received is retained and only the data with detected errors is retransmitted. As a result the technique improves the transmission efficiency over that which is possible with the Restart After Error Detection transmission protocol;

— it will be observed that all the curves in Figure 9.6 are of identical

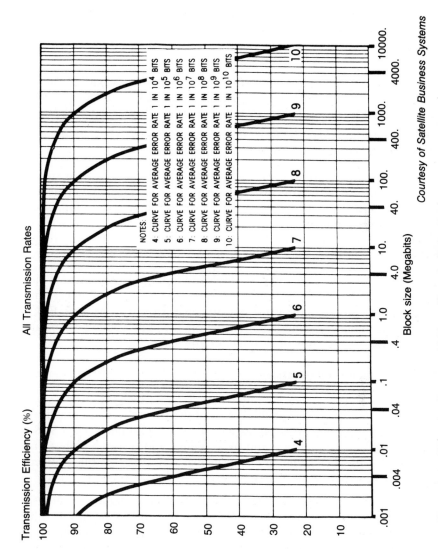

Transmission Efficiency (%)

All Transmission Rates

NOTES:

4. CURVE FOR AVERAGE ERROR RATE 1 IN 10^4 BITS
5. CURVE FOR AVERAGE ERROR RATE 1 IN 10^5 BITS
6. CURVE FOR AVERAGE ERROR RATE 1 IN 10^6 BITS
7. CURVE FOR AVERAGE ERROR RATE 1 IN 10^7 BITS
8. CURVE FOR AVERAGE ERROR RATE 1 IN 10^8 BITS
9. CURVE FOR AVERAGE ERROR RATE 1 IN 10^9 BITS
10. CURVE FOR AVERAGE ERROR RATE 1 IN 10^{10} BITS

Block size (Megabits)

Courtesy of Satellite Business Systems

Figure 9.6   Selective Retransmission: Transmission Efficiency vs Block Size

shape regardless of transmission rate. Therefore the efficiency is independent of transmission speed;

— inspection of Figure 9.7 reveals that transmission efficiencies of 90% or better, with a BER as high as 1 in 10^5 for all transmission rates of interest, can be achieved, providing that a block size no greater than 10,000 bits is used;

— of all the protocols examined, this one possesses the least sensitivity to changes in BER, for rates as high as 1 in 10^5, providing block sizes equal to or below 10,000 bits are used (see Figure 9.7).

— the observations regarding the performance criteria, window size, etc of the previous section also apply here;

— it is unnecessary to press the analysis any further, and it is sufficient to state the conclusion that Selective Retransmission meets all the criteria for handling the transfer of large data volumes at high data rates.

Forward Error Correction (FEC)

Forward Error Correction may be employed to improve the channel bit error rate (BER), and in conjunction with the protocol, will in certain circumstances improve the error performance. However, because FEC involves transmitting redundant information, the improvement will be at the expense of some loss of throughput efficiency.

Implementation Considerations

Commercial experience of satellite transmission at high bit rates is currently very limited, and in general most of the theoretical and practical investigations to date have been carried out by organisations such as SBS, and COMSAT Laboratories, and by academic and research bodies. In Europe this has been one of the principal objectives of the Project UNIVERSE experiment.

With a few exceptions satellite service suppliers including the PTTs have so far showed little interest in supplying packaged solutions. It will be apparent from the preceding discussions that to support high bit rates, and with block sizes which are close to the optimum, extensive buffer memory is required at the local ends of the link. And the implementation of Selective Retransmission also requires additional logic to keep track of

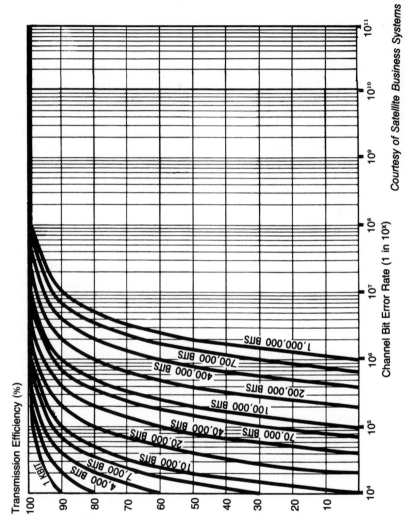

Figure 9.7 Selective Retransmission: Transmission Efficiency vs Channel Bit Error Rate (BER) (Blocksizes 1 Kbit – 1 Mbit)

the block sequences and to enable the sequence to be restored following a detected error.

In the USA, at least two products referred to as Delay Compensation Units have been developed to interface user devices to satellite links. One in particular is designed to support computer to computer or tape/disk to computer communication at transmission speeds in the range 56 Kbits/sec to 6.3 Mbits/sec. No information is currently available concerning either operational experience or cost. Nevertheless, the evidence suggests that the problem of delay for bulk transmission at high speeds is in principle solvable, the big question being: "At what price, and can it be commercially justified?". The answer lies somewhere in the immediate future, but one view is that – as is often the case – the leading edge multinationals, with their actual or potentially high traffic volumes and the technical and financial resources, will be amongst the first organisations to venture into this area.

Conclusions and Recommendations

Our extended discussions of the data delay problem has aimed to place it in perspective, and to show that, despite some of the adverse publicity, a great deal can be done to achieve a performance comparable with, and probably better than, that currently achievable with analogue terrestrial circuits. It is useful to summarise the principal conclusions and recommendations.

In implementing a satellite link, *it should not be treated as a direct substitute for a terrestrial circuit.* Such an approach may be successful if the appropriate protocols are employed together with any other modifications that may be necessary. Otherwise, significant reductions in performance may be expected.

The Space Segment Access Methods which are currently employed are not wholly suitable for random access from widely dispersed terminals generating bursty traffic.

The answer lies in the development and introduction of access techniques which take maximum advantage of the satellite's broadcast capability. The techniques which are currently being explored are sometimes referred to as Packet Broadcasting. (It should be noted that although information may be transmitted in a packet format, the techniques should not be confused with Packet Switching. The two are radically different.) Examples of the techniques include: ALOHA type procedures; Packet

Radio; and Burst Reservation systems.

With *non-ARQ terminals* (ie terminals operating in teletype compatible mode at speeds of the order of 30 characters/sec), full-duplex operation (echoplexing) is totally unacceptable. Input and echo time per character approximates to 706 milliseconds compared with say 20-50 milliseconds on terrestrial circuits. Half-duplex operation over a satellite link increases the response time per message by 640 milliseconds compared with terrestrial. There appear to have been few major complaints from users operating in this fashion.

With *ARQ terminals,* block-by-block protocols such as Binary Synchronous should be avoided over satellite links. Some form of continuous protocol should be employed, such as Restart After Error or Selective Retransmission.

Polling end-to-end across satellite circuits should be avoided because of its inefficiencies. Some form of remote polling confined to the terrestrial network is likely to provide an acceptable alternative. Some increase in transmission speed relative to what would be acceptable on a terrestrial network may be advisable.

With *data link control protocols* a continuous protocol should be employed in all situations. At speeds below 100 Kbits/sec and a BER of 1 in 10^6 or better, an efficiency greater than 90% is achievable with Restart After Error versions of continuous protocols such as HDLC, SDLC, etc (see Table 9.2).

With increasing BER and higher transmission speeds, there is a rapid decline in throughput efficiency. In these situations, a selective rejection protocol with its increased costs will be needed to achieve a satisfactory throughput.

Adjustments may be needed to the flow control parameters, buffer sizes, etc of Link Control Protocols and distributed network architectures such as SNA. Time-outs will need to be adjusted to the satellite environment.

The choice of window, block and buffer sizes are likely to be critical. For example, whilst Restart After Error will provide satisfactory performance at 9.6 Kbits/sec with a window size of 7 (see Figure 9.4), higher BERs and/or speeds will require this to be increased beyond this value to maintain a comparable throughput efficiency.

10 The Role and Application of Satellite Communications

SATELLITE COMMUNICATIONS IN CONTEXT

For much of its history, satellite communications has enjoyed – what seemed at the time – immense power and versatility unrivalled by traditional terrestrial communications services. In particular, a satellite communications system offered a transmission capacity over long distances, unobtainable by current earthbound technology. Moreover, on account of the broadcast property and its ability to transcend physical obstacles, satellite systems offered a geographical flexibility impossible to realise in finitely configured terrestrial transmission networks.

We have seen (in Chapter 3) how these advantages were progressively exploited: the initial successful application to international telephony; the distribution of TV programmes to local distribution networks; the supply of telecommunications services to inaccessible regions and to the less well developed countries of the world; and the introduction of Direct Broadcast TV (DBS) and Small-Dish Business Services.

All of this has been achieved over a time-span during which the nature of terrestrial transmission services in most countries changed only gradually, and was firmly tied to the provision of telephone services. In fact, outside the USA, with its more competitive telecommunications environment, the widespread exploitation of satellite communications has been constrained less by technological factors and more by political and legislative conditions governing their utilisation and accessibility.

The capabilities of terrestrial services and regulatory constraints are just two, albeit important, factors influencing the introduction and rate of growth of satellite services. Others include social and economic conditions, accidents of geography, and so on.

In order to understand the role of satellite services in Europe, how these will develop, and the costs and benefits of employing them for business purposes, it is important to appreciate the characteristics of the European telecommunications environment.

Competition from Terrestrial Services

The dramatic successes of satellite communications have encouraged the belief that it would become a major competitor with terrestrial communications – and might even supplant it. However, such a view underestimated the developments which were to occur in terrestrial transmission technology.

As in most other advanced countries, the European Telecommunications Administrations are either planning or implementing network modernisation programmes. The eventual aim is to replace the existing analogue networks by a wholly digital one employing optical fibre transmission and computer controlled switching exchanges. The networks will support services offering the following facilities:

— point-to-point leased circuits operating at speeds of 2.4 – 64 Kbits/sec and 2 or 8 Mbits/sec upwards;

— in the form of ISDN, switched services offering an end-to-end switched digital path between telephone terminals. In its simplest form this will provide dual multiplexed channels of 64 Kbits/sec and 8 Kbits/sec at the subscriber's telephone termination.

Not only do these services offer high bit rates, but with the introduction of fibre optic technology, the costs of terrestrial communications are now falling more rapidly than those of satellite communications. Progress varies between countries, but the UK has now reached a fairly advanced stage. Plans to install an optical fibre cross-channel link are under consideration, although there has been no firm decision. However, a transatlantic link (TAT8) is scheduled to come into operation in 1988.

It is clear that satellite services will face increasing competition from the new terrestrial services.

Geographical Factors

Excluding pan-European distances, domestic distances in European countries tend to be relatively short – which tends to reduce the cost benefits derived from distance insensitivity. Also, the major markets and user populations are geographically concentrated, and are already well supported by extensive terrestrial networks.

These factors have significant implications for the economics of earth station distribution and space segment access. In order to be competitive, the access costs have also to be reduced below what might be considered acceptable in a country such as the USA. It is possible to achieve significant savings in two ways. Because of the smaller coverage areas it is practicable to employ narrower, more concentrated, beams with a higher gain at the satellite, enabling smaller lower cost earth stations to be employed. And the existence of an extensive terrestrial network, coupled with relatively short terrestrial earth station access circuits, suggests that significant savings would be obtained by encouraging intensive use through shared or community earth stations.

There is one other implication of the shorter distances. For terrestrial routes, not only the costs but also transmission impairments such as distortion, noise and bit error rates depend on distance. However, a 500 km European link is often far better than the international reference circuits of the CCITT to which the quality standards of satellite communication systems have also been aligned to date. Although it could be argued that the transmission quality required for a given service ought to be distance-independent, national satellite links have sometimes had to be upgraded to a transmission quality in excess of international standards in order to gain market acceptance.

Social and Economic Factors

Telecommunications generally, and satellite services in particular, are an important instrument of policy in underdeveloped countries. In conjunction with factors such as a non-existent or rudimentary telecommunications infrastructure, an arid terrain and large distances, satellite services have been relatively easy to justify. In Europe, these conditions do not generally apply. On the other hand, the more advanced stage of economic development implies not only a correspondingly well developed terrestrial communications infrastructure, but also a receptive market for more sophisticated and novel services such as home interactive services, specialised business services, etc, as distinct from basic telecommunications services.

The Role of Television

Although our concern is primarily with business communications, television requires consideration for two reasons. First of all, together with telephony, it has supplied the basic financial underpinning for satellite services. It is now the most cost-effective way of distributing TV to local distribution nodes, and utilising transponder capacity for TV and telephony continues to provide the quickest route to financial break-even for satellite operators or owners.

A second reason is the recognition (first of all in connection with PRESTEL) that the domestic TV receiver, as well as providing entertainment, constitutes a display terminal that can be utilised for receiving and accessing a wide range of interactive services, either independently of or in association with personal computers. The introduction and market acceptance of non-entertainment services would open up a vast potential market for satellite services, either down-linked via cable systems (BSS) or by DBS. The latter would provide only a one-way service, although a telephone channel could be used for the return path.

Cable TV is off to a good start in Europe but has a long way to go before it achieves a penetration comparable with that in the USA. Also, satellite TV is the subject of intense debate involving such issues as the relative merits of Satellite/Cable versus DBS, and the viability of Pay TV. Nevertheless, despite the uncertainties, there is sufficient evidence to indicate that in Europe, as in the US, TV revenue will be an important factor in satellite economics.

Political Factors

The countries of Europe, with their cultural and linguistic differences, are a far less homogeneous grouping than is the United States, and this manifests itself in a number of ways. For example, countries vary in their attitudes towards the organisation and regulation of telecommunications and in respect of the powers which they exercise over their PTTs. There are wide variations across Europe in respect of advertising standards for TV and the content of programmes beamed into the various countries. For various reasons, governments tend also to be sensitive concerning "spill over" from other countries' satellites.

Problems associated with security, privacy, transborder dataflows, etc – which are likely to be exacerbated by the expansion of satellite services

– have yet to assume significant proportions, although such problems will almost inevitably arise in the future. Finally, cultural differences and differing priorities and attitudes increase the difficulty of reaching common agreement on such issues as service facilities and tariffs.

The Regulatory Environment

For strategic, social and economic reasons, the supply and operation of telecommunications services have, for most of their history, been subjected to some form of regulation and supervision both at governmental and international levels. In the United States, the policy has been to allow the marketplace to regulate itself through competition but within a broad regulatory framework. In contrast, countries in the rest of the world have traditionally relied upon their respective governments to determine the policies which govern their telecommunications systems. In these countries the responsibility for the supply of telecommunications products and services has been entrusted to a single national body enjoying monopoly powers. During the early evolution of telephony, several factors helped to cement the growth of national telecommunications monopolies. These are the most important: economies of scale and standardisation benefits deriving from a single supply source; and the priority attached by governments to ensuring the rapid spread and universality of the telephone. However, over the last decade or so there has been a growing recognition amongst industry, users, and some governments that, although monopoly supply may at one time have contributed significantly towards the development of the Public Telephone Network, in some important respects it is now having the opposite effect. Telephony no longer enjoys quite the supremacy that it once enjoyed; technology convergence is blurring the boundaries between computing and communications, and between information in all its forms; and the traditional regulatory frameworks are seen to exercise a constraining influence on the more widespread and effective exploitation of the new information processing technologies.

Consequently, there has for some time been growing pressure for the removal or reduction of regulatory constraints – nationally, internationally and within Europe. These pressures are now starting to bear fruit. For example, the United Kingdom is firmly embarked on this road, telecommunications in the US is undergoing a radical upheaval with the break-up of AT&T and the divestiture of the Bell operating companies, and the

European Commission is pressing for realistic and greater competition throughout the EEC.

Although the present regulatory frameworks were largely created during the evolution of the terrestrial telephone networks, they have also influenced the growth of satellite communications, particularly in the US relative to Europe.

Because both the historical development and the organisation and supply of services in most European countries differ significantly from the North American pattern, they will inevitably influence the way in which European satellite services will develop, at least in the short to medium term. Accordingly we shall now review the implications of the present regulatory environment in Europe.

Europe and the USA

In countries (eg the US) where telecommunications has an essentially horizontal structure, the growth of national satellite networks has been favoured by the fact that they were operated by independent carriers. This is not so in Europe. With the possible exception of the United Kingdom – where Mercury communications has gained approval to offer a competitive transatlantic service – satellite services will in general be operated by the same administrations which operate the terrestrial services.

In the now more liberalised environment of the United Kingdom, BT faces competition from the proposed Mercury service. However, until Mercury achieves a significant share of the market, BT's control of a highly developed terrestrial network, together with its long established overseas links, inevitably means that it will continue to exercise a dominating influence for some time to come.

Service Tariffs

Administrations have traditionally based their tariffs for terrestrial services not on the actual cost of supply of each distinct service, but on a cost-averaging process. Thus, for telephony, the charges for long distance calls are used to subsidise the charges for local calls, although the costs of local plant may be perhaps four times the cost of long-distance circuits. This was a major factor in the early expansion of US domestic satellite services, making it possible for the operators to under-cut the prices charged by terrestrial carriers on long-distance circuits.

Although the evidence is by no means easy to quantify, there are strong indications that National Administrations have tended to apply similar tariffing policies to international satellite services as to long distance terrestrial services. For example, INTELSAT levies uniform charges worldwide for use of the space segment by participating countries, but the tariffs charged to consumers by their respective Administrations may reflect a mark-up of 100% or more on the INTELSAT charges.

In fact, according to INTELSAT 90% of the local charge for an international circuit is attributable to the cost of gaining access to the space segment, so that only 10% of the price of an international telephone call goes to INTELSAT. Nor in their international charges do the PTTs distinguish between traffic going by satellite and traffic going by undersea cable. Therefore, because of the cost and price averaging which occurs, and the absence of detailed information, it is by no means evident that the distance-independent savings of the satellite route are fully reflected in the price which is charged.

Space Segment Access Arrangements

Until recently, access to the International Satellite Space Segment was restricted to National Gateway Earth Stations owned and operated by the Administrations of INTELSAT member countries (COMSAT in the US). This is now in the process of changing. INTELSAT and the UK Government have given approval to Mercury Communications to access the INTELSAT transatlantic segment from their own earth stations.

Whilst this chapter was in preparation, the FCC made a ruling which is intended to open up access to International Services by other US domestic satellite operations carriers besides COMSAT. However, it remains to be seen what policies will be adopted throughout Europe generally, since the INTELSAT member countries can make their own decisions in this area.

As regards access to domestic and regional satellite services in Europe, National Administrations which enjoy a monopoly are in a strong position to dictate access arrangements, and possibly restrict competition.

Satellite Services and Space Segment Competition

At the present time, INTELSAT holds the monopoly over the supply of international services, and the member country signatories are enjoined by the INTELSAT constitution "not to engage in any activity (meaning entering into competition with INTELSAT) harmful to its interests". And over the years, countries have obeyed this dictum, if not to the letter, at least in spirit, and a variety of stratagems have been employed to ensure that protocol is satisfied. Thus, some regional satellite services lease capacity to INTELSAT, and then perhaps lease back for their own use. However, the organisation has not been able to turn back the tide, and regional satellite services have begun to proliferate. There is now growing pressure for additional international services to be introduced in competition with INTELSAT. Several US companies have announced their intentions of filing applications with the FCC to offer such services, specifically on the busy transatlantic route. This whole subject is fraught with agonising problems for the US Administration.

Any decisions which are taken are not solely of domestic significance, but have foreign policy implications as well. INTELSAT represents a notable achievement in international cooperation – in striking contrast to the discussions which plague the wider political arena.

The INTELSAT system provides the same service at the same costs to nations with very high traffic demands and to nations which require only thin-route applications. It also provides domestic telecommunications services for countries which would not be able to otherwise afford them. Indeed it is doubtful whether, but for INTELSAT, the worldwide communications system would ever have materialised in the form that we see it today.

Service Integration

The vertical organisation of transmission services, which characterises much of Europe, facilitates the integration of satellite and terrestrial services. In conjunction with the smaller land areas this is likely to encourage the growth of shared or community earth stations and easier and widespread access. This is in sharp contrast to the USA where the supply of both satellite and terrestrial services is highly fragmented, and access to one or other of the seventeen or so domestic satellites may be partly determined by the attitudes and policies of the competing terrestrial carriers.

THE COMPLEMENTARY ROLE OF SATELLITE COMMUNICATIONS

The historical overview in Chapter 3 recorded the changing perceptions of the role and application of satellite communications up to the present time. The most prominent feature is the progressive broadening of its role so that it now encompasses:

In terms of geographical coverage

— global communications;

— regional and domestic communications;

— maritime/mobile communications.

In terms of services and traffic

— international, regional, etc trunk services;

— telephony, TV distribution, data transmission;

— small-dish customer-accessible business services.

At the same time we have also seen in previous discussions how the course of development and exploitation has been influenced and will continue to be influenced by a number of factors, including:

— geographical and political factors;

— local and international regulatory considerations;

— the relative rates of progress with terrestrial network modernisation and the growth of competition from enhanced terrestrial services;

— further advances in satellite and terrestrial transmission technology;

— competition for geostationary orbit and the frequency spectrum, and the extent to which their capacity limitations can be overcome and their utilisation improved.

Because of the combined and complex impact of these various factors both in space and time, it is perhaps simplistic to attempt to define a precise universally agreed role for satellite communications. And it would be virtually impossible to forecast with any accuracy how that role might evolve in the future.

However, at present it can be confidently asserted that satellite services *complement* terrestrial services. They may provide a cost-effective alternative to existing terrestrial services; they offer facilities not currently available terrestrially; and they possess inherent capabilities which would be physically impracticable to supply by terrestrial means.

Table 10.1 lists features in respect of which satellite systems complement terrestrial services. A number of these are mainly of interest to PTTs and service suppliers, and only of indirect interest to consumers, but there are others which are of direct relevance to business users. Most of the items in the table are self-explanatory, but we shall briefly comment on the more important ones.

Supply Flexibility

Where rapid development of a communications infrastructure is required, area coverage can be achieved quickly and more cost-effectively than by any other means. Satellites bridge large distances and inaccessible areas at costs independent of distance and the characteristics of the intermediate terrain.

Even within Europe there will be consumers in the less populated areas who are unable to obtain adequate services by terrestrial means. The North Sea oil platforms provide a good example, and in fact were the first locations in Europe to have direct connection to terrestrial services via satellite links.

The ability to supply mobile users wherever they may be, of which the INMARSAT system is currently the prime example, opens up a vast range of future possibilities, particularly if the size and cost of earth terminals continue to decline.

In conjunction with small-dish earth stations this flexibility offers important benefits to the business user:

— earth terminals can be installed on or adjacent to customer's premises;

— mounted on trailers they enable all the benefits of wideband digital transmission to be provided:

— on a temporary basis;

— for emergency communications, or to meet a short-term demand;

Feature	Comments
Supply Flexibility	
Difficult terrain	Most cost-effective and possibly only practicable route
Underdeveloped regions	Most cost-effective and possibly only practicable route
Early supply/access	Shorter lead time by using mobile earth terminals
Temporary supply	Facilitated by mobile earth terminals
Dispersed users	
Dynamically changing user populations	
Access by mobile users	Eg cellular radio; vehicle, shipboard or aircraft mounted earth terminals
Capacity Flexibility	
Redistribution/load balancing	Flexible redistribution of capacity between users and between satellite and terrestrial
Relatively elastic	Relatively straightforward to provide additional increments of capacity – launch another satellite
Performance Characteristics	
Global wideband capability	Currently available in principle
Digital transmission	Global operation will be achieved earlier for satellites than for terrestrial
Interconnection Flexibility	
Infinite	In principle infinite connectivity between all points within coverage area
Flexible network reconfiguration	Relatively easy to add or subtract earth terminals
Distance-Insensitive Costs	
Point-to-point circuits	Economic over long distances, but in general the break-even point is variable
Point-to-multipoint circuits	May be economic over the shorter distance where these are links of a multipoint or broadcast configuration

Table 10.1 Satellite Systems Features Complementing those of Terrestrial Systems

— in remote areas which are either not currently served by equivalent terrestrial services, or where the latter cannot be economically justified.

Early Availability of Wideband Digital Transmission Services

Satellite services provide end-to-end wideband digital links independently of the varying stages of development of national and international digital networks.

High-capacity users who can justify a private earth terminal avoid the lengthy and costly terrestrial access connections to international gateway stations. And customers sharing a community earth station can gain access through a relatively short dedicated circuit.

Thus, satellite services will make all the benefits of wideband digital communications available long before the world's terrestrial networks are fully converted and modernised.

Capacity Flexibility

One great advantage of satellite communications is in the flexible utilisation of a given total capacity it permits at any site, in terms of different services and bit rates. In the case of national and international services, for example, it provides capacity which can be allocated more flexibly than with terrestrial networks. It can be readily utilised to handle traffic overloads on the terrestrial network, and the ability to reallocate traffic between alternative services provides a powerful guarantee of continuity of service. It is also worth noting that an incremental increase in satellite capacity can be achieved more quickly than by laying additional undersea cables, simply by launching another satellite.

High Interconnection Potential

This derives from the broadcast feature. In principle, all points within the footprint or coverage area can interconnect with one another. The concept of a discrete and predetermined physical network of conductors, characteristic of terrestrial communications networks, is no longer applicable.

The broadcast property is particularly suited to point-to-multipoint or broadcast applications, whether the transmission is unidirectional or asymmetrical to receive-only earth stations, or two-way symmetric. Examples of such applications in business are videoconferencing, electronic message services, information distribution, document distribution and remote printing.

Appropriate space segment access methods such as TDMA which allow flexible use of capacity are a prerequisite for widespread business use. Furthermore, the ability to redistribute transmission capacity within a TDMA environment introduces a new element into the design and operation of distributed computing networks. It enables transmission capacity to be allocated and shared between users in a far more flexible manner than is possible in private terrestrial networks.

Distance-Insensitive Costs

The actual costs of providing a satellite link are largely independent of distance, and this is an important element in the economics of using a satellite link in preference to a terrestrial service.

The crucial figure, when all the costs have been taken into account, is the break-even distance, or the distance beyond which it becomes cheaper to use a satellite link rather than an equivalent terrestrial circuit.

Two cases need to be considered: the point-to-point circuit; and the multipoint circuit, comprising a central node connected to a number of destinations by point-to-point links in the broadcast coverage area. Depending upon the tariff structure, it is reasonable to suppose that in the latter case the average break-even distance is somewhat shorter than in the single point-to-point situation.

We return to this when we consider some numerical examples in Chapter 11.

APPLICATIONS AND APPLICATION OPPORTUNITIES

Currently, organisations such as ITU, EUTELSAT and the PTTs when discussing satellite services tend to distinguish between business or "new" services, and those services which were introduced primarily to augment traditional terrestrial services. The latter include classical telephony, broadcast TV and mobile services. However, the distinction is becoming increasingly artificial and to some extent confusing. Resulting from the convergence of technologies, service boundaries are becoming increasingly blurred; and digitisation is enabling all types of information to share the same digital paths. And, in addition, the geographical reach of communications is being progressively extended, enabling information to be disseminated ever more widely.

The ubiquitous TV receiver now stands in as a home computer display or as a PRESTEL terminal, and if current expectations are fulfilled, cable TV will offer access to a wide range of interactive services such as home banking, shopping, education, and so on. It is evident that none of these on-going developments can be entirely ignored by the business community.

The role of the supplier is also changing. Whereas traditionally the telecommunications administrations concentrated on the supply of basic services such as PSTN, private circuits and TELEX, they are now broadening their offerings to include advanced value added services such as packet switching, TELETEX, video conferencing, etc.

With these qualifications, we commence the following review of the established and traditional applications before considering the more novel opportunities which fall within the broad "new services" category.

Traditional and Miscellaneous

Classical Telephony

With the conversion to digital transmission, the distinction between speech and other forms of information is becoming increasingly difficult to maintain. Nevertheless the demand for telephone traffic continues to grow at rates very much higher than that of the world economy as a whole, whilst the growth rate for data and other non-voice traffic is higher still, although in absolute terms its contribution is still relatively small. For example, in South East Asia, annual traffic growth rates around 20% prevail, similar to the rates experienced in the EEC and America until the economic recession. The present depressed rate of their economies has reduced this – probably only temporarily – to around 15%.

Therefore, satellite communications will continue to make a significant contribution to meeting world telephony demands in the years to come. However, as the world's terrestrial and submarine cables are progressively replaced by optical fibre, a number of authorities anticipate some migration of traffic from satellite to terrestrial networks.

Tables 10.2 and 10.3 give statistics on the growth and usage of the INTELSAT network, compared with the cable network, and illustrate how cable and satellite systems have evolved in parallel to serve largely complementary roles.

Year ending 31st December	1965	1975	1982
Number of cable routes in service	57	115	208
Total installed route mileage (thousands)	55	111	163
Number of installed circuits	4200	48400	159700
Average provisioned circuit capacity per route in service	74	420	768

Courtesy of Standard Telephones & Cables

Table 10.2   World Submarine Cable Network 1965-82

Year ending 31st December	1965	1975	1982
Number of earth stations in service	5	115	230
Number of point-to-point traffic paths established	1	406	1088
Number of activated full-time circuits	75	6682	29730
Average activated half-circuits per earth station	30	116	259
Average activated circuits per traffic path	75	16	27

Courtesy of Standard Telephones & Cables

Table 10.3   INTELSAT Network 1965-82

It can be seen that cable systems are preferred when the traffic density on the route is high, with an average capacity of 768 circuits per route. In contrast, satellite systems are preferred, and are often the only practicable or economic choice, when the traffic density is low. This is reflected in the average use of less than 300 half circuits/earth station and fewer than 30 active circuits/traffic route. (A half circuit corresponds to either an up-link, or a down-link.)

At the present time there is international agreement between the PTTs and international carriers that traffic should be subdivided approximately 50/50 between satellite and terrestrial circuits. Table 10.4 and Figure 10.1 illustrate the historical and projected growth in INTELSAT capacity.

The European regional satellite system operated by EUTELSAT was originally conceived as a reinforcement of the European terrestrial network for public telecommunications services and to satisfy the requirements for TV programme exchanges by the European Broadcasting Union (EBU). It is planned to route between one third and one half of European traffic with a circuit length exceeding 800 km via satellite links.

Satellite	Date	Capacity in telephone circuits or TV channels
INTELSAT I	1965	240 telephony or 1 TV
INTELSAT II	1967	240 telephony or 1 TV
INTELSAT III	1968	1 500 telephony or 4 TV
INTELSAT IV	1971	3 000 telephony + 2 TV
INTELSAT IV-A	1976	6 000 telephony + 2 TV
INTELSAT V	1980	12 500 telephony + 2 TV
INTELSAT V-A	1984/5	14 000 telephony + 2 TV
INTELSAT VI	1986	up to 40 000 telephony + 4 TV

Reproduced from reference 10 in Bibliography by
Courtesy of the International Telecommunication Union

Table 10.4   Growth in Capacity of the INTELSAT System

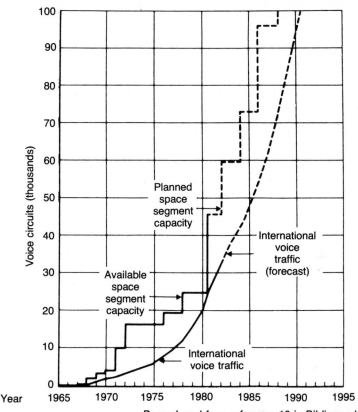

Reproduced from reference 10 in Bibliography by
Courtesy of the International Telecommunication Union

Figure 10.1 INTELSAT: Traffic Growth and Available Capacity

Broadcast TV Distribution (BSS)

Satellite transmission is a very cost-effective technique for distributing TV signals, either to cable head-ends or to microwave local distribution networks.

The distribution of TV programmes by satellite to cable networks, hotels, schools, etc is a booming business in N America. In contrast, the expansion of cable networks in Europe has, until recently, been limited to a few small countries like Belgium, Switzerland and the Netherlands, but it now seems to have got off to a good start in several large countries such as France, the Federal Republic of Germany and the UK.

In Europe the OTS satellite has been used regularly to relay pro-grammes of the French TDF and of the UK Satellite TV Co. These transmissions, which started on an experimental basis, very quickly assumed a pre-operational character. Satellite TV's programme, now called SKY Channel, has been transferred to ECS–1 which also transmits TV5, a programme jointly produced by the French, Belgian and Swiss broadcasting organisations. Both SKY Channel and TV5 are on the air every evening and distributed by more and more cable networks in Europe. Quite recently they started to become available through cable systems in the UK.

The basis of the United Kingdom Government's declared strategy in encouraging the development of Cable TV is that the revenues derived from entertainment TV in the earlier stages could supply the funding to develop a variety of interactive services for domestic use. If and when such services do arrive, their acceptance is likely to be facilitated by a number of factors: the vast increase in the ownership of personal comput-ers, and the apparently insatiable appetite for them; the availability of the domestic TV receiver as a display terminal, the level and distribution of personal disposable incomes; and social and cultural attitudes.

In addition to entertainment we could expect programmes of an increasingly specialised character to be offered, providing information on a variety of topics such as finance, sports, hobbies, etc. The possibilities are endless and limited principally by cost, market acceptance and ingenuity. But the combination of high picture quality with interactive capabilities enables a wide range of new services to be supplied, amongst which the following have already been identified (some of them, such as electronic newspapers, home banking and software distribution, are already either being evaluated or are operating on a pilot basis):

— improved definition TV;

— multilingual dubbing and subtitling;

— stereophonic sound;

— videotex and electronic mail;

— electronic newspapers;

— home banking and shopping;

— distribution of software and data files to personal computers.

These developments will create new opportunities for information providers and will have widespread implications for the conduct and structure of business and commerce generally.

Direct Broadcast TV (DBS)

If one excludes the large number of pirate earth stations in the US there are currently no operational DBS services worldwide, although there has been considerable experimental activity in the US and Europe. However, full services are planned to come into operation in the US from 1985 onwards, and the provision of DBS services is under active consideration in several European countries, including the UK. Services can be expected to become available via both the EUTELSAT system and domestic satellites. The precise role and economics of DBS, in particular its relationship to Broadcast Services (BSS), is the subject of considerable debate at present. Part of the reason is that DBS can be regarded as both competing with and complementing BSS services, but there is also considerable uncertainty concerning its financial viability.

In its complementary role DBS provides an attractive solution to distributing TV to locations in areas which are either not served by cable or community antennae systems, or where it would be uneconomic to do so.

DBS would be able to supply an equivalent range of receive-only services to BSS, but at present it is difficult to see how interactive services could be provided at an economically attractive price. The current opinion is that the broadcast channel would serve to transmit information to the user, whilst the latter's instructions and messages could be handled adequately by a telephone channel.

Mobile Communications

The INMARSAT maritime communications system affords the prime example of the application of satellites to locations which are on the move. And with the next generation of satellites which are now being planned, it is proposed to extend access to aircraft as well as to ships. Thus, like ships, aircraft in flight would be able to establish direct connection with the terrestrial networks.

Within the terrestrial framework, radio paging systems have been available for a number of years, and these enable people to similarly access the public telephone network. However, their coverage has been limited geographically, their expansion has been hampered by scarcity of frequency spectrum, and their quality leaves much to be desired. The recently developed cellular radio technique employs frequency re-use, and enables both the capacity to be increased, and the geographical coverage to be extended. It also offers a considerable improvement in service quality.

However, two problems remain. Cellular networks will be installed as a matter of priority in urban areas where the demand is likely to be greatest, and this will leave large regions without any service at all.

Also designs and frequency bands proposed for cellular networks vary from country to country, so that user terminals will be governed by a multiplicity of different standards. Thus, a traveller whose vehicle is equipped with radio telephone will be unable to communicate as soon as he leaves his own area, either because the country he enters is served by a different system, or because no service is provided.

A satellite system could usefully plug the gaps in the coverage of the terrestrial networks and at the same time reconcile the different national standards. The European Space Agency (ESA) is currently carrying out studies with the objective of defining the satellite channel characteristics required to serve very small radio receivers which can be carried on all types of vehicle.

Commercial and Allied Applications

In the USA the most common application of satellite services in business is as a cost-effective replacement for leased circuits in private terrestrial networks. A typical private network comprises satellite links for the long-distance trunks, with cable and/or microwave circuits employed for shorter distances and for local distribution. The majority use of satellite services has been for telephony either solely or accompanied by data transmission, the latter generally at speeds up to 56 Kbits/sec. Until recently, the market for specialised business services, which enable the wideband and multipoint capabilities to be effectively exploited, has been slow to develop. (In Chapter 3 we noted that the SBS service which was aimed primarily at this market has had to resort to more traditional offerings because of the relatively slow growth.) Nevertheless, there has been (by European standards) substantial experience of applications such as video conferencing, information distribution and remote printing and publishing, which either require wideband transmission or can benefit from high data rates.

Within Europe most of the national administrations and EUTELSAT have made commitments to offer business services, although partly for reasons given earlier in this chapter the market is difficult to assess. A small number of mainly US-based multinational organisations have been using satellite services into Europe for several years, employing the 56 Kbits/sec SCPC channels on the Atlantic region INTELSAT satellite, and there have been several European-sponsored experiments to investigate the benefits of satellite communications at high bit rates. The following discussion of typical satellite business applications draws upon the US experience and these European studies.

Data Transmission and File Transfer

For services in the ECS coverage area under present plans, priority is being given to access at speeds of 64 Kbits/sec upwards, in multiples of 64 Kbits/sec, to 2 Mbits/sec. This speed range is considered to be an essential requirement. At the lower end it is compatible with the terrestrial network PCM hierarchy, and at the upper end it provides the capacity to handle image and video conferencing applications. However, the high minimum bit rate of 64 Kbits/sec effectively excludes users with low data transmission requirements, or applications which can be handled adequately by the traditional lower-speed services, ie 1200, 2400, 4096 Kbits/sec, unless the user can multiplex several lower-speed channels onto the one link, or otherwise concentrate traffic from several sources.

At the present time a commonly expressed view in Europe is that users with low-speed data transmission requirements can in many cases be better served by terrestrial means. On the other hand it should be noted that on the French TELECOM series of satellites, it is proposed to offer a range of speeds which starts at 2.4 Kbits/sec. However, the longer-term future for the traditional low-speed services is closely bound up with the development of terrestrial digital services, in particular ISDN, the impact on bandwidth costs, and the market implications. The UK pilot ISDN service provides an important pointer, since the simplest form of access provides dual speeds of 64 and 8 Kbits/sec, and the tariff is *independent of transmission speeds,* in contrast to the traditional lower-speed services. The extent to which in the longer term the lower terrestrial speeds are superseded by higher speeds is a matter for speculation. What is certain at the present time is that satellite services fill an important gap, offering high bit rates and digital transmission independently of either distance or the locations served.

Prior to the introduction of satellite business services and fibre optic technology, the application of telecommunications has been conditioned largely by the capabilities of the existing terrestrial services. The speeds achievable extended up to perhaps 19.2 Kbits/sec with a large gap until either 48 Kbits/sec or 56 Kbits/sec was reached, and these were by no means universally available. Furthermore, the sheer logistics and costs involved in acquiring the higher speed circuits made them prohibitive except for all but the large corporations and governmental organisations. The prospect of an increasingly widespread penetration of services offering high bit rates at an economic price, whether provided by satellite or by the new terrestrial services, opens up a new world of novel opportunities which would have been scarcely conceivable a few years ago, so rapid has been the revolution in technology. This dramatic shift in capabilities is inevitable, requiring corresponding efforts on the part of suppliers and users to appreciate the implications and exploit the opportunities.

The availability of and ready access to high transmission speeds and some fairly straightforward calculations will quickly cause us to re-evaluate our perceptions concerning the feasibility of transferring large volumes of data. Table 10.5 shows the elapsed times for transmitting files of various sizes as a function of the transmission speed.

The table does not contain equivalent data for speeds below 56 Kbits/sec, but it can be seen that, even at this relatively high speed, transmission times for sufficiently large files may be unacceptably long. For instance, the time taken to transmit the contents of a 625 Megabyte disk approximates to about 28 hours.

A number of applications which could exploit or benefit from the high transmission capacity of satellite links have already been implemented or are being investigated. Perhaps the first notable use was for transmitting the large volumes of data arising from seismic exploration. Another is the distribution and processing of data collected by the meteorological and earth resources (LANDSAT) satellites.

In the commercial sphere, satellite transmission has been employed for several years for the simultaneous printing of newspapers and periodicals at remote locations. Other applications include the distribution and

TRANSMISSION RATE	DATA FILE SIZE (bits)					
	10^7	10^8	10^9	10^{10}	2×10^{10}	3×10^{10}
56 Kbits/sec	3.3 m	33.1 m	5.5 h	2.3 d	4.6 d	6.9 d
112 Kbits/sec	1.7 m	16.5 m	2.8 h	1.2 d	2.3 d	3.4 d
224 Kbits/sec	49.6 s	8.3 m	1.4 h	13.8 h	1.2 d	1.7 d
448 Kbits/sec	24.8 s	4.1 m	41.3 m	6.9 h	13.8 h	20.7 h
1.544 Mbits/sec	7.2 s	1.2 m	12.0 m	2.0 h	4.0 h	6.0 h
3.152 Mbits/sec	3.5 s	35.3 s	5.9 m	58.8 m	2.0 h	2.9 h
6.312 Mbits/sec	1.8 s	17.6 s	2.9 m	29.3 m	58.6 m	1.5 h

Courtesy of Satellite Business Systems

Legend: s = seconds
          m = minutes
          h = hours
          d = days
Assumed throughput efficiency = 90%

Table 10.5 Transmission Time versus Transmission Rate for Various Data File Sizes

maintenance of large databases, high-speed facsimile and document distribution.

In Europe several projects have been carried out to explore how satellite channels could be used to speed up the exchange and distribution of large quantities of data. For example, the primary objective of the STELLA project involving CERN (the Centre for European Nuclear Research at Geneva) and a number of Universities and research establishments was to investigate how wideband satellite links could be employed to speed up the processing of the vast amounts of data generated by particle accelerator experiments. Computer analysis of the data requires several days to complete, and this invariably must be carried out at the home establishments of the participating scientists, with much time lost in travelling and sending magnetic tapes backwards and forwards. In the first phase, the experiment was limited to transferring between two computers bulk data stored on magnetic tape. In the second phase of the experiment more flexibility was introduced into the system by interconnecting local area networks rather than isolated computers, thereby providing multiple access capability, and resulting in greater efficiency and autonomy of operations.

The European Space Agency (ESA) has also demonstrated on a number of occasions the transmission of documents by high-speed facsimile. The demonstrations established that very high quality copies of A4 size could be produced at a rate of one every four seconds. The experiments also illustrated the potential value of another very promising application, namely the distribution of documents by libraries and information services. This application is the subject of the APOLLO project which is currently in the planning stage, and will be described later on.

Remote Printing and Publishing

Satellite links, with their high transmission capacity and broadcast capability are assuming growing importance in the printing industry. For newspapers the principal advantages are:

— they enable the pages to be edited and composed at a central location, and then transmitted simultaneously to a number of geographically remote production plants;

— newspapers and magazines with a wide circulation can reach readers in different countries at the same time as in the country of origin, rather than two or three days later;

— because deadlines are considerably shortened, actual production can be delayed to allow the incorporation of the latest news. At the same time it still allows local editions to be prepared or material of local interest to be inserted at the remote site.

For weekly illustrated magazines with their longer production cycles, it is perhaps not immediately apparent that the technique can offer major advantages. The major problem for this class of publication is colour, without which a colour magazine would lose much of its appeal. Each colour page is printed from four colour separation plates, and their preparation is both complex, time-consuming and expensive. However the competitive pressures are such that if a current affairs magazine is to maintain its market share it must be able to reproduce pictures of recent events shortly after they have happened. This means that in practice both the text and the graphics would have to be set almost as the presses were rolling.

Until recently the preparation of colour graphics has not been able to take advantage of the rapidly improving techniques for text preparation which are being progressively introduced into newspaper production. This problem has now been overcome with the development of high resolution colour graphics composing equipment, which can be simply described as the graphics designer's equivalent of the word processor.

In the process an A4 photograph or graphic image is broken down into 1024×1024 individual cells, and on completion of editing the digitised image is then stored away on the composing unit's 300 Megabyte disk drive. Subsequently the stored images can be retrieved and input to other equipment which converts the image to film, ready for plate preparation. This operation can be performed either locally, or the image can be transmitted to a remote location and the film production carried out there.

However, remote production entails the transmission of very high data volumes, since a *single* A4 colour separation requires *30 Megabytes* when digitised at a resolution suitable for high-quality print production. (This contrasts with a maximum of 6 Kbytes for a page of word processed text.)

Currently, the system now in operation transmits at rates of 56 Kbits/sec, and in order to achieve an acceptable turnround, the data is compressed before transmission to effect a reduction in transmission volume of 8:1. Otherwise a typical mix of colour and text could take about $1\frac{1}{4}$ hours for a single A4 page.

These applications were pioneered in the USA – although much of the advanced high-resolution editing and scanning equipment now coming into use is of UK design and manufacture. A number of newspapers and periodicals are now being printed by this method, both in domestic and international editions. Examples include: *The Wall Street Journal* (14 sites), *Time* magazine (9 sites) and *USA Today* (22 sites).

In the UK the first experimental investigations took place in 1981, and involved a link between the *Financial Times* in London and their Frankfurt operations, using a speed of 154 Kbits/sec. Then, in early 1983, the *Economist* and the *Manchester Guardian Weekly* commenced printing of their North American editions in this way. Prior to using the satellite service, the *Economist* employed couriers to take the plates for the entire magazine by Concorde to New York and then by road to the printers in Connecticut. The journey took eight hours and was at the mercy of fog and strikes. By satellite it takes about one hour to transmit the full colour cover, and the black and white pages take about 4 minutes each, the total transmission time amounting to about seven hours. It is worth observing that some of the time saved comes from avoiding US Customs, since courier carried film counts as a commodity requiring customs clearance.

For the *Economist* the costs are claimed to be about the same, and for the *Manchester Guardian Weekly* slightly more expensive, than the previous procedures. In both cases the significant benefit is the time saving which allows the print run to commence between 8 hours and half a day earlier, and thus beat their competitors to the newspaper and magazine stands.

A number of the printing and publishing houses currently employing the technique are now actively investigating its extension to other regions of the world, particularly the Middle and Far East, and this seems set to become a major growth area for satellite communications.

Document Delivery and Distribution

There now exist many databases which anyone equipped with a simple interactive terminal can access and interrogate in order to retrieve useful information, and information database services such as those supplied by EURONET DIANE, TYMNET, etc are in widespread use. However, whilst the services enable the user to consult bibliographic abstracts covering a particular topic in a few minutes, obtaining copies of documents and articles of interest is an entirely different matter. Acquiring and distributing copies may take several days or even weeks, and besides imposing severe clerical burdens on libraries and other information providers, the delay clearly reduces the effectiveness of the information service. Therefore a system which could speed up the transmission of documents and reduce the delivery time to a few hours or minutes would be of considerable interest to many organisations and individual users.

According to the result of an investigation sponsored by the European Commission the number of requests for documents processed by all the libraries in Europe is well in excess of 10 million a year. For example, the British Library Lending Division alone handles more than 10,000 requests per day. It was estimated that if all the requested documents were to be delivered electronically, the total amount of information to be transmitted in digital form would amount to about 100 Gigabits (100×10^9) per day. This could be comfortably handled at an average transmission rate of 1 Mbit/sec which, when considered globally, constitutes a modest load for a satellite system. Such a transmission speed would enable documents to be delivered at a rate of the order of 1 page/second, which could not be currently achieved economically on a large scale with the existing terrestrial networks.

These considerations have prompted the European Space Agency and the European Commission to establish, in conjunction with EUTELSAT and other organisations, the APOLLO project – the objective of which is to design and develop a pilot document tranmission service. In the initial phase, it is proposed to establish three electronic libraries at Luxembourg, Frascati in Italy, and the British Library at Boston Spa. The receiving earth stations will be either individual or shared depending upon local requirements. The terminal equipment will comprise a variety of elements enabling information to be displayed on a screen, printed on paper or stored on magnetic media. A user will interrogate the database through a terrestrial circuit in the customary manner, and place requests for the delivery of selected documents. These will then be retrieved and transmitted over the satellite link.

In addition to handling requests for specific documents, it is also intended that the APOLLO system should be available for automatic distribution of European Community Documents to national and international organisations within the community. The present plan is for the system to come into operation in early 1986 using the second or third flight models of the ECS satellite.

Although the growing use of electronic storage, transmission and display methods can be expected to reduce the use of the printed word, there will continue to be many situations in business and commerce where information must be retained in documentary form, whether to satisfy legal requirements, or merely for convenience. Examples include: conventional commercial documents such as invoices and the like; contract notes recording stock exchange and commodity dealings; and legal contracts at various stages of drafting and negotiation. Therefore, document transmission whether directly from a stored image or by facsimile is a potentially important business application.

Video Conferencing

National video conferencing services employing terrestrial transmission networks have been around for some time in several countries; examples are the Bell Picture Phone service in the US, and CONFRAVISION in the UK. But, due to a combination of high transmission costs and the sheer inconvenience of participants having to travel to their nearest studio, services such as CONFRAVISION have failed to achieve a significant impact.

Because of the wideband transmission requirements, and the need to interconnect a number of widely dispersed locations, video conferencing would seem to be a natural candidate for satellite communications. However, despite the oil and energy crises, and the efforts of organisations to develop and promote satellite video conferencing services, there was continuing market resistance. Although some of this was undoubtedly due to user inertia, costs constituted a major obstacle. Transmission costs remained high, because, until it became possible to transmit video signals digitally and compress them into a narrower bandwidth, conventional TV broadcasting technology had to be employed and this required a full 36 MegaHertz transponder. When digital compression did become available, this, together with lower-cost, small-dish earth stations, prompted organisations like SBS to develop and promote services which could be accessed directly from customer premises using private earth stations and studios. However, costs continued to place it beyond the reach of all but those organisations who were able to justify the earth station and studio costs. Nevertheless several US corporations have established video conferencing facilities but many have held back to await the arrival of lower-cost systems.

A video conference can be organised in three different settings:

— as a "special event" in which delegates attend their nearest hotel or other centre, the latter being equipped to handle a large number of participants;

— a small number of participants attend a local shared studio;

— a small number of participants utilise a studio located on company premises.

Of these three modes of operation, the first is now a fast growing industry in the US, and is spreading to Europe. For example, companies such as Ford of Detroit employ the technique for their annual sales conventions. It enables them to reach audiences as large as 20,000, resulting in major savings in travel and associated costs, and improving the dissemination and exchange of information.

A number of American and international hotel groups have equipped their premises to handle video conferencing and others are planning to do so. The Intercontinental Hotel in London which is linked into the SBS INTELMET service is an example. The majority of conference applications in this category still employ broadcast TV and one-way video via TVRO earth terminals, a telephone return path being used for audience questions, etc.

Within the last 2-3 years, service and equipment suppliers have been directing attention to the provision of lower-cost company-located facilities; the aim being to avoid the high cost of specially constructed studios, and to enable video conferencing services to be accessed via low-cost terminals, which can in principle be transported from office to office. A number of these products are now coming onto the market.

In the United Kingdom, video conferencing services to Canada via satellite have been opened, and similar services to the USA, Europe and elsewhere are in negotiation or planned. It is also planned to introduce domestic video conferencing services using the terrestrial network. These will be marketed under the title of VIDEOSTREAM, which is now undergoing pilot trials. When fully operational, this will offer a switched service between customer premises and will have connections to international video conferencing services.

Information Distribution

A prominent application in the US is the dissemination of information to receive-only earth stations. This ranges from news of all kinds to specialised financial information. The operators of these services include news agencies, offshoots of conglomerates and information service companies.

In the UK, Exchange Telegraph (EXTEL) recently carried out trials involving the broadcasting of information to nine of their branch offices throughout the country.

Tele-education

This has much in common with video conferencing, and is a fertile field of exploration at present. In an earlier chapter we noted the pioneering use of satellite communications for this purpose in Canada and Alaska. Another early pioneer of tele-education, although not employing satellite links, has been the Open University in the UK. Currently, there is an experiment underway involving the educational authorities in the West Country, and extensive investigations are in progress in the USA, Europe and Japan. A number of commercial organisations in the USA are actively exploring its potential for sales training and training of customer's staff.

Whilst its primary aim is to present educational material remotely, it can take a number of forms. It can be handled live or be prerecorded; it can be conducted by speech only, or speech plus video; and may be supported by local or remote computer software and databases.

11 Satellite Business Services in Europe and UK

INTRODUCTION

Having discussed broadly the role and opportunities of satellite communications, it is now appropriate to focus attention on UK satellite service plans and opportunities. We briefly review service plans and intentions, and then make a tentative assessment of the role and economic benefits of small-dish services as they might affect the business community.

Satellite business services are being actively marketed (1984 heralds the commencement of European and UK satellite services). However, there are a number of reasons why it is at present difficult to make an objective assessment of the costs and benefits for business:

— with the exception of the North American Video Conferencing Service launched in 1984, small-dish services are only just emerging;

— services are still in process of being defined in terms of their facilities and associated tariffs;

— the absence of significant user experience and the limited published tariff information preclude an accurate assessment of costs and benefits at the present time;

— whereas in the USA there has been something like five years experience of small-dish services, for European suppliers and users alike it is an entirely new venture. And, as was discussed in Chapter 10, the USA experience does not necessarily translate directly to the European environment.

The foregoing are essentially short-term considerations, but in the medium to long term, other factors will shape the evolution and growth of services. These include: the enhanced capabilities of the next generation of satellites which are planned to be launched during the next five to ten years; and the effect of changing cost and tariff differentials between terrestrial and satellite services.

For these reasons, the following discussion, together with any conclusions which are advanced, derive from currently available information on service plans and costs.

UK SERVICE PLANS

The two UK service providers are BT and Mercury Communications, the latter having been granted approval by the UK Government and INTELSAT to offer private circuits on the transatlantic route, and to operate earth stations in the UK.

BTI has so far announced plans for small-dish services to North America, Western Europe and UK offshore locations, and these are currently being promoted under the SatStream banner. These services are discussed below.

SatStream North America

This came into service in early 1984, operating initially between London and Toronto. The service will be progressively expanded during the year by the installation of additional earth terminals in Eastern Canada and the UK, with extension to the United States by the end of the year.

SatStream Europe

This service is planned to open in the Autumn of 1984, when the ECS–2 satellite becomes available for commercial use.

SatStream Offshore

This is intended to offer a specially designed service to the UK offshore industry, linking offshore production platforms with Aberdeen. A successful experiment was run in 1982/3, using small dishes at BTI's Mormond Hill radio station and AMOCO's Montrose Alpha platform. Data, voice and telegraph circuits were provided from August 1982 to March

Service Area	Satellite	Operational Date	Service Provider	Operator or Sponsoring Organisation
Europe	ECS–2 onwards	1984	BT	EUTELSAT
	TELECOM–1	1984 (?)	BT	French PTT/ EUTELSAT
	L–SAT series	1986 (?)	BT	EUTELSAT
Europe, N America	UNISAT	1986 (?)	BT	United Satellites consortium
(1) North America	INTELSAT (AOR) (2)	Current	BT	INTELSAT
(1) North America	INTELSAT (AOR)	1984 (?)	Mercury Communications	INTELSAT
USA	INTELSAT and a US carrier	1984	BT	INTELSAT and a US carrier
Middle and Far East	?	?	BT	Under consideration
(3) Global	INTELSAT System	1986	BT/Mercury (?)	INTELSAT

Notes: (1) Capacity available for private business use but only accessible through National Gateway Earth Stations.

(2) AOR = Atlantic Ocean Region Satellite.

(3) INTELSAT Small-Dish Business Services.

Table 11.1   Satellites Offering Business Services

1983. In September 1982, slow-scan TV facsimile, teleconferencing and electronic mail were demonstrated.

Under whatever description the services are marketed, they will utilise a combination of satellites, depending upon the geographical coverage, the space segment access method and performance requirements. The current service plans in terms of the satellites which will be used are summarised in Table 11.1 (domestic, European region and international satellites are all represented).

Within Europe services will be provided by the ECS satellites, starting with the second flight model of ECS (ECS–2), and the TELECOM–1 satellite. Services will also subsequently be offered on the L–SAT series.

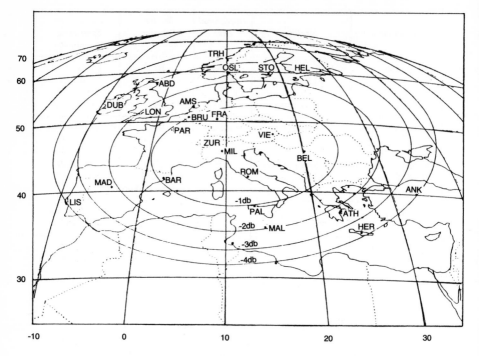

Figure 11.1   ECS Business Service Coverage

The French TELECOM–1 domestic satellite is being made available for European wide service by agreement between the French PTT and EUTELSAT. Figures 11.1 and 11.2 illustrate the coverage areas for ECS and TELECOM–1 respectively, and Figure 11.3 the initial disposition of the large dish or national Gateway earth stations.

The proposed UK UNISAT domestic satellite carries a telecommunications payload and an antenna configuration which will enable services to be provided between the UK, North America and Europe, and, subject to regulatory agreement, also between the USA and Canada.

INTELSAT has for some time offered capacity for private business use, and the SatStream North America service utilises the Atlantic Ocean Region (AOR) satellites. However, in 1983 INTELSAT announced their intention to provide business services on a global basis. The Model V satellites operate in both the C and Ku frequency bands, and global coverage is to be achieved by satellite on-board cross connections bet-

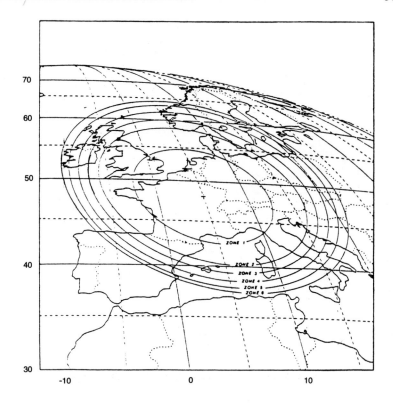

Figure 11.2 TELECOM-1 Business Service Coverage

ween the Ku spot beams covering major commercial centres and the hemispheric or zone beams operating in the lower 6/4 GHz frequency band. This will be realised with the introduction of the specially modified V-B satellites in 1986. In the meantime BTI has plans for early implementation of cross-connection facilities involving other US carriers to give expanded coverage in the USA. It is also working on plans to introduce small-dish satellite links with the Middle and Far East.

It will be observed from this summary of current plans that only INTELSAT will be able to offer a truly global service. However, how long they will maintain this unique advantage will depend upon several factors including: competitive pressures; successful continuation of their international carrying monopoly; the rate of expansion of Regional Sys-

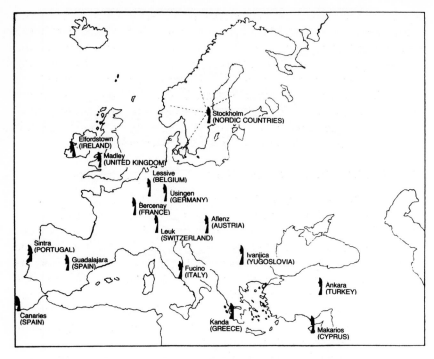

Figure 11.3   Initial Disposition of ECS Earth Stations

tems and the technical and political feasibility of linking these together and thus bypassing the INTELSAT system.

It should be noted that all of the above satellites provide multiple services, and that in addition to business services, a substantial portion of transponder capacity is dedicated or planned to be used for telephony or TV.

SERVICE PRESENTATION

In the UK, services are currently being defined and marketed in a form which largely conceals the particular satellite which may be used, together with its technicalities. Service offerings are currently taking two forms: either as a bearer service, or to support a specific service. In the former, the customer purchases transmission capacity of specified amounts, and performance characteristics, to use as he wishes, whereas in the latter a service package to support a specific application, such as video conferencing, is supplied. Customer access can either be through a shared or community earth station, or by private earth station. For marketing purposes, BTI are treating the two types of access rather differently. Shared access is being presented as a *structured service* with defined speeds, access circuits and associated tariffs, whereas private earth station access is considered to be an *unstructured service*. In the latter case, because of the potentially wide variation in customer requirements, it is proposed to treat each individual application separately.

SPACE SEGMENT ACCESS AND CAPACITY ASSIGNMENT

With the exception of business services which use TELECOM–1, according to present published information, services offered on the other satellites in Table 11.1 will operate in SCPC and/or FDMA modes, at least initially. However, TELECOM–1 will offer TDMA access, and the future provision of TDMA on other satellites is under consideration. As we noted in Chapter 8, the access method has a significant bearing on the structure of the service, how it is supplied and the flexibility it offers to the user. Thus, services operating under SCPC or FDMA will in general be offered in channels of capacities which are customer pre-assigned, and with access being scheduled in advance.

Without the flexibility offered by TDMA, a customer would need to define and evaluate his requirements within these constraints. For example, if he wished to obtain the equivalent of a private terrestrial network, this would have to be constructed out of pre-assigned point-to-point or point-to-multipoint satellite links.

SATSTREAM EUROPE FACILITIES

The following are the facilities which BTI are proposing to offer on the SatStream Europe service.

Earth Station Access

— initially in London and at Aberdeen, but later in major commercial centres throughout the UK. For users who wish to access a shared earth station the local access circuit to the user's premises will be either KiloStream or MegaStream. Customers will have the option of purchasing sole use earth stations either from BTI or from approved independent suppliers.

Transmission Rates

Although these were listed in Chapter 7 they are repeated here for convenience.

— 64 Kbits/sec and selected multiples up to 1.920 Mbits/sec;

— 2.048 Mbits/sec which can be used for data, but also to support video conferencing applications;

— several other options will be available derived from submultiplexing the 64 Kbits/sec channel. They include: five 9600 bits/sec channels; and a 32 Kbits/sec voice channel, accompanied by one or more lower speed data channels.

Space Segment Channel Configurations and Utilisation Arrangements

The following circuit and channel configurations are offered:

— one-way channel or two-way circuit;

— one-way or two-way point-to-point;

— one-way point-to-multipoint (broadcast).

Capacity can be supplied and accessed under the following alternative arrangements:

— part-time basis;

— full-time basis;

— switched connections;

— provision of "hot line" or access at short notice is under consideration.

ECONOMICS OF SATELLITE BUSINESS SERVICES

In the US, users have been able to achieve major cost savings through employing long-distance satellite, rather than terrestrial, circuits. Now that business satellite communication services are becoming available in Europe, the question of major interest and concern to the business community is whether such cost savings are applicable in the European environment, and the conditions under which they are achievable. It has, until recently, been virtually impossible to make any well-informed, quantitative assessments. In particular, BTI has not published the full tariffs for SatStream, although "indicative" tariffs which are subject to revision have been disclosed.

The cost comparisons and conclusions presented here draw substantially upon a recent (and, to the best of our knowledge, the only) published study quoted in reference 8 in the Bibliography. This study carried out cost comparisons between terrestrial and satellite services for applications in the US, within mainland UK and within Europe as a whole. The study considers several typical point-to-point and multipoint examples, and computes the economic break-even points between satellite and terrestrial services.

(The reader should note that, because of imperfect information regarding SatStream and terrestrial service tariffs, earth station costs, etc, the calculations require a number of assumptions and estimates to be made. For example, the indicative tariffs which are quoted are subject to confirmation.)

It is considered worthwhile to include quantitative examples: firstly, because they help to place UK satellite business services in a practical context; and secondly, because they might help the reader to gain some insight into the nature of the cost relationships and those factors which are of critical significance. For fuller details, the reader is recommended to read the original paper.

It is appropriate to start with an example which typifies US experience, and illustrates the relative economics of satellite and terrestrial services in the US environment.

A US Example

At the present time it is difficult to compare the costs of alternative service offerings in the US. Resulting from the divestiture of AT&T and the consequent restructuring of their tariffs, the rates of nearly all carriers are in a state of flux as they re-evaluate their competitive positions. For the purpose of the exercise AT&T rates filed with the FCC are used. The two AT&T services which are compared are the SKYNET 1.5 Mbits/sec satellite service, and the comparable ACCUNET T1 1.5 Mbits/sec digital terrestrial service.

ACCUNET can be accessed either from shared-use earth terminals which are available in a number of cities, or from dedicated terminals. AT&T do not however offer receive-only terminals; terminals can be transmit-only, or transmit and receive, but one or more receive-only channels can be added to a transmit earth station which customers either share or own.

The monthly rental costs for the two services are plotted against distance in Figure 11.4. This reveals that the break-even points between satellite and terrestrial point-to-point circuits in relation to distance and the earth terminal configuration are:

— shared earth terminal (one-way)   390 miles

— shared earth terminal (two-way)   880 miles

— sole use earth terminal (two-way)   1375 miles

It is instructive to apply these results to an example. A brokerage house with headquarters in Manhattan wishes to establish a corporate network linking their headquarters mainframe computer with eight trading offices

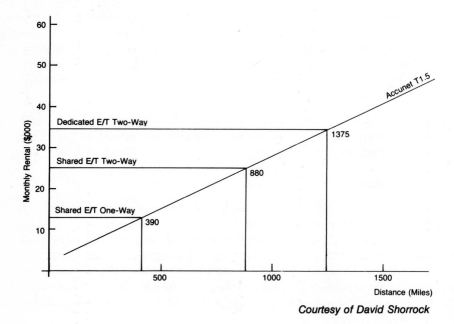

Courtesy of David Shorrock

**Figure 11.4   Distance Break-Even Points for Satellite vs
Terrestrial Circuits – USA**

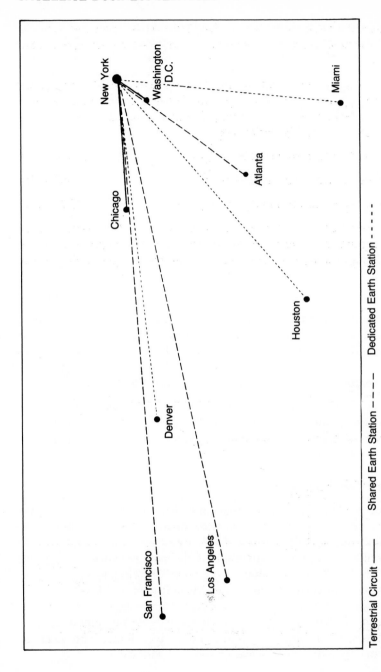

Terrestrial Circuit ——— Shared Earth Station – – – – Dedicated Earth Station - - - - -

Figure 11.5   A Mixed Satellite and Terrestrial User Network

scattered across the continental US (see Figure 11.5). However a shared earth terminal service is not available in all cities. The optimum configuration can be arrived at by applying the appropriate tariff tables and the results of the break-even analysis to the various permissible combinations. The resultant network, as shown in Figure 11.5, comprises both satellite and terrestrial links:

New York	— Washington DC	terrestrial
	— Chicago	terrestrial
	— San Francisco	shared earth terminal
	— Los Angeles	shared earth terminal
	— Atlanta	shared earth terminal
	— Denver	dedicated earth terminal
	— Houston	dedicated earth terminal
	— Miami	dedicated earth terminal

This mixed solution would offer a cost saving of over 60% compared with a solely terrestrial system. Such an approach is a common feature of the communications strategies employed by US corporations.

Satellite versus Terrestrial Services in Mainland UK

Within the UK mainland, SatStream will be in competition with the new digital terrestrial services which are progressively coming into operation. Before considering the economics of using SatStream we briefly review the terrestrial service plans.

UK Digital Terrestrial Services

The network modernisation strategy on which BT is currently engaged calls for a completely digital main transmission network by the early 1990s; and roughly half of this will employ optical fibre 140 Mbits/sec trunk circuits. The present coverage of the new network is shown in Figure 11.6. The X-Stream family of digital private circuits is based upon this network, and currently comprises MegaStream and KiloStream.

MegaStream. This is the highest capacity service offering rates of 2, 8, 14 or 34 Mbits/sec. Customers are connected to an exchange with access to the network by either microwave or specially laid cables.

Figure 11.6   Present Status of the UK Digital Transmission Network

KiloStream. This is a full-duplex synchronous transmission service offering rates of 2,400, 4,800, 9,600, 48K and 64K bits/second. It commenced in January 1983, and by December 1983 was available at over 200 main exchanges throughout the UK. Local access from customer premises is generally over the existing local metallic loops suitably conditioned for the service.

Tariffs

SatStream. BT have not published full tariffs for SatStream, although "indicative" tariffs have been disclosed (see Table 11.2). The tariffs apply to shared earth station use, and include the cost of a dedicated customer MegaStream or KiloStream access circuit up to 15 km from the earth station.

For sole-use earth stations, BT are not prepared to release tariff information, preferring to quote on an individual basis. Neither are they willing to quote tariffs for small-dish receive-only earth stations. However, the capital cost of a receive-only terminal with a 1.8 metre antenna suitable for working to the ECS–2 service is currently around £12,000, a

Service and Data Rate Shared Earth Station	To Europe	To North America
2-way 64 Kbits/sec	£ 30,000	£ 50,000
1-way 64 Kbits/sec	26,000	40,000
2-way, 1 hour per day, 64 Kbits/sec	12,000	23,000
2-way 1920 Kbits/sec	500,000	600,000
1-way 1920 Kbits/sec	483,000	565,000

Note: Costs shown are the annual rentals for UK end only.

Courtesy of David Shorrock

Table 11.2   British Telecom Indicative SatStream Tariffs

figure which is expected to halve in quantity production.

MegaStream. BT do not publish tariff information for MegaStream, preferring to quote on an individual customer basis. One paper (reference 8 in Bibliography) employs the following estimates for a 2 Mbits/sec circuit:

—   an annual rental of £152 per kilometre between exchange nodes, plus a £1,500 base charge;

—   connection charges of approximately £20/kilometre plus a £350 base charge;

—   the connection charge for the access circuit between the customer's premises and the nearest network access exchange is separate and depends upon the amount of engineering work necessary. Typically this could be between £15,000 and £80,000.

KiloStream. The KiloStream tariff for 64 Kbits/sec circuits in excess of 15 kilometres comprises:

—   an annual rental of £5/kilometre together with a base charge of £3,500;

—   there is an initial connection charge of £400/circuit.

Point-to-Point Circuits

It is clear that for applications involving transmission links restricted to the UK Mainland, SatStream costs must be compared with those of MegaStream and KiloStream. Table 11.3 has been compiled by taking the SatStream tariffs of Table 11.2 and computing the terrestrial circuit distances which these equate to, using the quoted and estimated terrestrial tariffs. The SatStream tariffs in Table 11.2 apply only to one end of the circuit, and must be doubled to arrive at the figures in Table 11.3.

Although a number of assumptions have been made, it is evident that there is a considerable divergence between the break-even distances for the US and the UK. The relevant figures for 2-way shared earth stations are:

Terrestrial Circuit	Break-even Distance
US T1 1.544 Mbits/sec	880 miles
UK 1.92 Mbits/sec	6,570 miles

Service and Data Rate	SatStream Annual Rental	KiloStream Equivalent Distance
2-way 64 Kbits/sec	£60,000	7,019 miles
1 way 64 Kbits/sec	£31,000*	3,416 miles
		MegaStream Equivalent Distance
2-way 1920 Kbits/sec	£1M	6,570 miles
1-way 1920 Kbits/sec	£503,000**	3,300 miles

Note: * assuming receive-only earth station cost of £5,000
** assuming receive-only earth station cost of £20,000

Courtesy of David Shorrock

Table 11.3   SatStream versus Terrestrial Point-to-Point Circuit Costs

Cost Comparison: Point-to-Multipoint

Because of the much greater circuit length, multipoint, star or broadcast applications – in which information is to be transmitted from a central point to a number of dispersed locations – are prime candidates for satellite communications.

The following example compares the costs of SatStream and Kilo-Stream for transmitting information from London to a number of provincial locations, whose mean distance from London is 200 miles. The information is transmitted at 64 Kbits/sec and is up-linked using a one-way channel to receive-only earth stations. The annual rental of a single KiloStream 64 Kbits/sec circuit for this mean distance is £5,110.

For a given mean distance, the critical parameters in a cost break-even analysis are: the receive-only earth station costs, and the number of receive-only earth station nodes. Using the transmit earth station rental given in Table 11.2, the number of receive-only nodes at which SatStream breaks even with KiloStream can be calculated for a range of assumed receive-only earth station rentals. The results are as follows:

Assumed Rental/annum	£2,000	£3,000	£4,000	£5,000
Number of Nodes	9	13	24	237

This shows that for an assumed receive-only earth station rental of £2,000, SatStream breaks even with KiloStream when the number of nodes is 9 or greater. However, for a rental of £5,000 the number of nodes needs to be 237 to break even and to recover the transmitting earth station costs. Thus, for receive-only earth station rentals in excess of £5,000 (or more exactly the KiloStream circuit rental of £5,110 considered in the example), SatStream fails to recover the transmitting earth station cost, and a KiloStream-based network works out more economic, as well as providing a full-duplex link to each node.

SatStream Services within Continental Europe

At present all terrestrial services between the UK and Europe involve a submarine cross-channel cable link which employs analogue technology. For this reason it is not possible to compare the costs of a digital Sat-Stream link to Europe with an equivalent terrestrial digital service as was possible for the US examples and those restricted to mainland UK. For terrestrial services the choice lies between M1020 circuits offering 9.6 Kbits/sec, and 48 KHz wideband circuits offering 64 Kbits/sec. Table 11.4 shows the annual rentals of the alternative satellite and terrestrial services between London and four European countries. It should be noted that in some instances the tariff of the remote end of the terrestrial circuit, or of the other half circuit of the satellite link, has been estimated.

Therefore a user wishing to transmit information at 64 Kbits/sec by terrestrial means between, say, London and the capital cities of these countries would either have to employ multiple 9.6 Kbits/sec circuits or a 48 KHz circuit suitably equipped to handle 64 Kbits/sec. With the data employed, the table demonstrates that a SatStream 64 Kbits/sec circuit offers significant cost savings over the terrestrial wideband equivalent.

	Analogue Circuits		SatStream Circuits	
	Annual Rental		Annual Rental	
From UK To	M1020 (9.6 Kbits/sec)	48 KHz Wideband (64 Kbits/sec)	64 Kbits/sec	2 Mbits/sec
FRANCE	£11,538	£89,512*	£60,000**	£1.0M**
GERMANY	22,771	165,240	60,000**	1.0M**
BELGIUM	12,185	103,140	60,000**	1.0M**
SWEDEN	21,119	154,801	81,624	1.36M

Note: * includes estimated tariff of French PTT
      ** includes estimated tariff of distant PTT

Courtesy of David Shorrock

Table 11.4   Satellite and Terrestrial Service Tariffs within Europe

CONCLUSIONS

We have observed elsewhere that the commonest business use of satellite services in the US has been the cost-effective replacement of terrestrial services. This is a direct consequence of the favourable conditions which obtain and which include: a large land area and lengthy transmission distances; a competitive supply environment; and a low-distance, break-even point (less than 400 miles) between satellite and terrestrial services.

Within the UK and the wider European context, these conditions either do not apply or apply to a lesser extent. The small UK land area and short distances imply that even though satellite services were to be competitively priced, and the tariffs accurately reflected the cost of provision, the potential savings from distance-independent costs will be less than those achievable over substantially longer distances.

As to the competitive environment, although there are two satellite service providers in the shape of BT and Mercury, both the services and the market are at an early stage of development, and there has been insufficient time for competitive factors to operate and manifest themselves. Satellite services in general also face strong competition from the new digital terrestrial services, particularly in the UK and to an increasing extent in Europe. Owing mainly to the use of fibre-optic transmission, terrestrial transmission costs are currently falling at a faster rate than are satellite costs.

The specimen calculations indicate that the break-even distance between satellite and digital terrestrial services for point-to-point circuits in mainland UK varies between 3000 and 7000 miles depending upon the data rates. And for broadcast applications, satellite services only cost in for networks with a large number of nodes, and this assumes a receive-only capability in the remote earth stations. Even allowing for tariff and estimating uncertainties, the calculations suggest that tariffs would need to be reduced substantially for satellite services to provide an economic alternative to terrestrial services for many business applications in mainland UK.

However, this conclusion does not invalidate the use of satellite services in a complementary role. A user may require either permanent or temporary access to digital wideband services in an area not currently served by equivalent terrestrial services. And in such circumstances, the suppliers may consider it more expedient or economic to provide the service by satellite.

For applications across Europe, SatStream has to be compared with traditional analogue terrestrial circuits, and here satellite services do cost in. This is likely to continue to apply in the short-to-medium term, at least until there is substantial penetration of digital terrestrial services within Europe.

Although no quantitative evidence has been advanced here, these results may be expected to apply in the wider international context. As in Europe, digital satellite services have to be compared with terrestrial and submarine circuits which are predominantly analogue; the benefits deriving from distance independence should be proportionally greater; and satellite services also offer wideband digital services to users in advance of the introduction and interconnection of national digital networks.

A final question concerns the economics of private small-dish earth station access versus shared earth station access. Whilst earth station costs are forecast to reduce substantially over the next few years, European PTTs, and also British Telecom, are in a favourable position to integrate satellite circuits into their national networks, thereby providing universal access. In these circumstances, unless earth station costs fall dramatically, there will be little incentive for organisations to install their own earth stations. However, multinational organisations with heavy traffic loads, or applications requiring wideband digital services, may be better placed to economically justify a private earth station. For such organisations, which may already employ satellite trunks in their global networks, the incremental cost of adding additional earth stations will be small in relation to the resulting benefits and their total telecommunication budgets.

Epilogue:

Satellite Communications in Perspective

This book has been prepared at a time when virtually all areas of tele-communications are experiencing a rapid and unprecedented rate of development and growth – a process which may be expected to continue in the years ahead. Satellite communications is just one of the tech-nologies contributing to this dramatically changing scene. The other main ingredients are: fibre optics, digital transmission, cable TV, cellular radio, local area networks, and computer-controlled PABXs.

Satellite communications was soon seen as having a unique power and versatility, besides which the capabilities of traditional terrestrial tech-nology seemed quite puny. And this was the prevailing view until fibre optics started to be employed in the terrestrial cable networks. This development, having both cost and performance implications, is inevit-ably causing a re-evaluation of the role of satellite communications. At one extreme it is argued that, as fibre optic technology increases its penetration of terrestrial networks, the role of satellites will diminish, resulting in significant migration of traffic to terrestrial services.

At the other extreme those with a vested interest in satellite technology will point to the major technical advances being planned. Where does this leave us? In the short-to-medium term, there is healthy competition which will result in major benefits whichever way the argument goes. But how the issue will be resolved is impossible to say. None of these new transmission tools and services has reached its technological limits or has yet been fully exploited.

We have seen (in Chapter 10) that the new digital terrestrial services can, in terms of cost and performance, already compete effectively with satellite links. However, although the optical fibre circuits currently being

317

installed operate at 140 Mbits/sec, there are some operational systems which transmit at 400 Mbits/sec, and in the research laboratory, transmissions at frequencies of 1 million MHz (1,000 Ghz) have been achieved.

Apart from the potential for further increases in performance, most authorities also confidently predict a continuing reduction in terrestrial transmission costs. Any comparison of terrestrial and satellite communications must also take account of the three significant limitations inherent in the latter: propagation delay; finite orbital capacity; and frequency spectrum crowding.

This book has provided glimpses of ongoing developments in satellite communications technology. These include: increased size and power of satellite systems; increased frequency re-use to conserve the frequency spectrum; the move towards 30/20 GHz and higher frequencies; and the use of multiple spot beams. For example, although the frequency constraint cannot be eliminated, saturation can be postponed, and some authorities are now claiming that the technology is available to enable a fifteen fold re-use of the frequency band to be achieved.

The projected space platform is also expected to have a major impact on orbital capacity and frequency re-use. This will enable large antennae – as big as a football field – to be mounted in space and supplied with sufficient power to direct large numbers of concentrated spot beams to earth. As an example, one of a number of design studies by NASA proposes an antenna of over 200ft diameter generating 10,000 beams, which would be capable of being received by small personalised antennae. There is obvious scope for the development of satellite communications technology in the future.

Two areas where satellite communication has a major advantage, and where it is likely to remain supreme for the foreseeable future, are: the supply of services to sparse populations and over hostile terrains; and for mobile communications, particularly over extensive tracts of land, sea, and airspace. These are either technically or economically impracticable to cover by terrestrial cable or microwave systems. Mobile services are now at the point where major expansion can be expected over the next few years.

The capabilities of some of the new services, which are now at an advanced planning stage, are certainly breathtaking in their implications. Three organisations in the USA recently announced their intentions to provide advanced mobile services, accessible by small personalised transceivers and covering the continental United States and other parts of North America. These proposals are currently being considered by the FCC. Such services offer two-way communication either for speech, messages or data, or for all three, and might eventually support as many as 12,500 to 30,000 two-way channels.

In some of these cases the term "mobile" may be too specific and perhaps the description "personal satellite service" would be much more appropriate, as it would encompass both the moving and the non-moving applications and serve the public at the individual level rather than on a community or corporate basis.

What for many people is likely to be the most remarkable aspects are that a personal satellite telephone might weigh no more than 5 pounds, and that telegraphic messages could be transmitted or received by a calculator-size, hand-held terminal powered by torch batteries. They are therefore eminently transportable, and could be carried on-board a car, boat or aircraft and in a brief-case or handbag.

In addition to establishing two-way contact with a remote location, two of the planned services also offer an additional capability. By employing two satellites in tandem, they not only establish radio contact but can also pinpoint the geographical location of the "paged" destination.

A wide range of potential uses are being promoted, and apart from conventional telephony and message services, these include such novel applications as: directional guidance and position reporting for vehicles, aircraft, etc; detection of emergencies and despatch of emergency services; supplementing air traffic control; avoidance of mid-air collision; and terrain avoidance for boats and aircraft. Thus, these enhanced mobile services provide a particularly powerful and effective solution to the problem of how to find and establish two-way communications wherever the individuals or organisations are situated.

A central theme remains the evolving role of satellite communications in relation to other emerging transmission technologies. To a large extent the various technologies have evolved independently of one another. It is therefore not surprising that technologies and services should overlap in capabilities, and lead to new competitive relationships. But, however the overall scenario evolves over the years ahead, the complex of technological factors, commercial competition and social forces will determine that satellite communications performs those jobs for which it is best suited.

A major task which lies ahead is to integrate all transmission services so that people, wherever they are, will be able to freely intercommunicate. This will not be achieved easily, nor within a short timescale, but the reader can be assured that this is a realistic goal, the long-term aim being to provide a worldwide communications infrastructure appropriate to the needs of the early part of the next century. When this has been achieved, the *global village* will be close to reality.

Bibliography

TELECOMMUNICATIONS AND DATA TRANSMISSION CONCEPTS

1 *NCC Handbook of Data Communications* (2nd Edition Edited by G B Bleazard), NCC Publications, 1982.

2 *Modems in Data Communications,* P R D Scott, NCC Publications, 1980.

3 *Computer Communications,* R Cole, MacMillan, 1982.

4 *Telecommunications: A Systems Approach,* G Smol, M P R Hamer and M T Hills, George Allen and Unwin, 1976.

SATELLITE COMMUNICATIONS

5 *Satellite Communications,* James Martin, Prentice-Hall Inc, 1978.

6 *Satellite Communications,* Proceedings of an International Conference, On-Line Publications Ltd, 1980.

7 *Satellite Communications,* Proceedings of an International Conference, On-Line Publications Ltd, 1984.

8 *User Economics of Satellite Communications,* David Shorrock (this paper is included in the preceding publication).

9 *Sixth International Conference on Digital Satellite Communications,* Phoenix, Arizona, 19-23 September 1983.

10 *Economic and Technical Impact of Implementing a Regional Satellite Network,* ITU, Geneva, 1983.

11 *Space Telecommunications* (special issue of *Telecom France* journal), French Ministry of Posts, 1983.

12 *Satellite Transmission Protocol for High Speed Data,* James L Owings, Satellite Business Systems.

13 *Satellite Communications,* British Telecom International, 1981.

14 *Notes on Satellite Communications Systems Engineering,* PSF Technical Seminars, Wilbur L Pritchard and Joseph A Sciulli.

15 *Satellite Communications,* Seminar Notes, Frost and Sullivan seminars on Satellite Communications, 1983.

16 Performance of High-Level Data Link Control in Satellite Communications, A K Kaul, *COMSAT Technical Review,* Volume 8, Number 1, 1978.

JOURNALS

17 *Satellite Communications,* monthly, Cardiff Publishing Company, Cardiff Communications Inc, 6530 South Yosemite St, Englewood, C080111 USA.

18 *Cable and Satellite Europe,* monthly, Cable and Satellite Magazine Ltd, 533 King's Road, London SW10 0TZ.

19 *International Journal of Satellite Communications,* quarterly, John Wiley & Sons Ltd, Baffins Lane, Chichester, Sussex PO19 1UD or 605 Third Avenue, New York, NY 10158.

Appendix A

A Note on Decibels

The decibel is a unit of measure employed by telecommunications engineers for comparing signal strengths. Unlike the watt which measures the absolute power of a signal, the decibel is defined in terms of the ratio of the respective powers of two signals. For example, the two signals being compared might be the input signal to a channel and the received signal at the other end, or the received signal and the received noise, ie, the signal-to-noise ratio. In general, a transmitted signal experiences various *losses* through attenuation, and also *gains* introduced by amplification. A central goal of the telecommunications engineer is to minimise the losses and to maximise the power of the received signal relative to the original transmitted signal. In those situations where the concern is primarily with relative signal strengths, the decibel is a more useful measure than the watt.

Historically the decibel was first employed as a unit for the measurement of sound levels in telephony. In this context it was found appropriate to use a logarithmic unit of measurement, because the response of the human ear is proportional to the logarithm of the sound energy, not to the energy itself. However, the decibel now has a far wider application and it is used to express, in addition to differences in noise intensity, such quantities as: losses in transmission, noise levels, and gain in amplification.

The decibel value is calculated as the logarithm (to base 10) of the ratio of the powers (in watts) of the two signals, multiplied by 10, ie:

$$\text{Number of decibels} = 10 \log_{10} \frac{\text{P1 watts}}{\text{P2 watts}}$$

If the ratio is greater than unity then the decibel figure will be positive, and if less than unity it will be negative. For example, if P2 is taken to be an input signal, and P1 the output signal, then the positive value signifies that P2 has experienced a gain in strength; conversely, the negative value indicates that it has suffered a loss.

Apart from its physical significance the decibel also has a valuable algebraic property in that, being logarithmic, losses or gains in signal strength can be added and subtracted. A numerical example (in Chapter 6) shows how this is utilised in calculating the Link Budget, from the

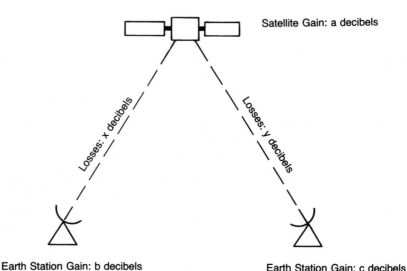

Satellite Gain: a decibels

Losses: x decibels

Losses: y decibels

Earth Station Gain: b decibels

Earth Station Gain: c decibels

Total Gain/Loss = (a+b+c−x−y) decibels

Figure A.1   Addition of Gains/Losses on a Satellite Link

summation of the various gains and losses which occur across the satellite link (see Figure A.1). The example also indicates that in cases where a signal lacks a corresponding signal value for comparison (for example, when the value for P1 is the initial transmitted power), for consistency the value of P2 is taken to be unity. The decibel value is then referred to in a unit of decibels/watt. For example: 10 watts = 10 decibels/watt, 100 watts = 20 decibels/watt, etc. With this convention all the gains and losses along the link are then fully additive.

Appendix B

Glossary of Terms

Amplifier   Device which increases the power of an electrical signal.

Amplitude   Loudness, strength or volume of a signal.

Amplitude Modulation   A modulation technique in which the amplitude of a signal is the characteristic which is varied to transmit intelligence.

Analogue Transmission   Transmission of a continuously variable signal (usual transmission method for telephone lines) as opposed to a discretely variable signal (as in digital transmission).

Antenna Gain   A directional antenna also functions as an amplifier, concentrating the power that is fed to it, and the radiation that it receives (see Gain).

Apogee   The point on an elliptical orbit where the satellite is furthest from the earth (cf Perigee).

Asynchronous   A transmission method where the signals employed may be of variable duration (eg Morse code).

Asynchronous Transmission   Transmission in which each character is preceded by a start signal and followed by one or more stop signals. A variable time interval can exist between characters. The start signal triggers off the receiver and, following receipt of the character, the stop signals switch off the receiver in readiness for receipt of the next character.

Attenuation

Decrease in magnitude of current, voltage or power of a signal in transmission (signal loss or fading).

Audio Frequencies

Frequencies that can be heard by the human ear when transmitted as sound waves. (Maximum human audibility range 16-20,000 Hz.)

Bandwidth

Range of frequencies available for signalling in a communications channel. The difference between the maximum and minimum frequencies on that channel.

Baseband Signal

In its narrow definition this refers to the transmission signal prior to its modulation onto a carrier. However, the signal received at an earth station from the terrestrial network may already have passed through one or more stages of modulation, prior to the modulation onto the radio frequency carrier performed in the earth station. It is customary to refer to this signal as baseband *relative to the earth station radio frequency modulation stage*.

Baud

Unit of discrete signalling speed per second: the modulation rate. Baud = bits per second when two-state signalling is used. Because of the baud rate limitations over speech circuits, multi-level signalling is used to obtain high bit rates on these circuits.

Bit

Abbreviation of binary digit. Signal element of transmission in binary notation either '0' (OFF) or '1' (ON).

Bit Error Rate (BER) The proportion of transmitted bits which are received in error.

Bits Per Second (bits/sec) Transmission rate possible on a circuit (bit rate).

Block A group of characters transmitted as a unit.

Buffer A storage device used to compensate for a difference in rate of flow of data, or time of occurrence of events, when transmitting from one device to another.

Carrier (Frequency) System A means of deriving several channels over a single link by causing each channel to modulate a different 'carrier' frequency and demodulating at the distant end to obtain the original signal.

Channel (Communication) An information transfer path. Channels are commonly derived by multiplexing and do not in general correspond to physical circuits.

Character Letter, figure, number, punctuation or other sign contained in a message; usually represented by one byte.

Check Bit or Check Character A bit or character associated with a character or a block for error detection purposes.

Circuit (2-Wire) A circuit having a pair of conductors over which both send and receive transmissions take place.

Circuit (4-Wire) A circuit where the send and receive directions each have a separate pair of conductors.

Circuit Switching Conventional interconnection where a two-way fixed bandwidth circuit is allocated exclusively to the parties concerned for the duration of the call.

Common Carrier

National organisation which has the authority/responsibility to provide public telecommunications services in its own country.

Demand Assignment Multiple Access (DAMA)

Whereas the FDMA, TDMA and SCPC multiple access systems, in their basic form, allocate fixed amounts of capacity to users, the incorporation of DAMA principles enables the allocated capacities to be varied dynamically according to individual demands. It is not employed in traditional FDMA systems, but is used in conjunction with SCPC – SPADE being an example – and TDMA systems.

Demodulation

Process whereby the original signal is recovered from a modulated carrier. The reverse of modulation.

Differential Modulation

A modulation technique where the coding options are related to the previously received signal, eg phase angle of the received signal related to the phase angle of the preceding signal.

Digital Transmission

In digital transmission, information is represented by discrete pulses or signal levels.

Duplex Channel

A channel capable of transmitting in both directions simultaneously.

Duplex Transmission

Transmission which takes place in both directions simultaneously.

Earth Segment

Refers to the earth station portion of a satellite link.

Equalisation

Means of improving circuit quality by equalising different distortions.

Equivalent Isotropic Radiated Power (EIRP)

This is a measure of the radiated power of an antenna. It is derived from the antenna gain and the power fed to the antenna.

Error Detection and Feedback System (ARQ)

A system employing an error detection routine and so arranged that the receiver, on identifying an error, initiates a request for retransmission of the corrupted data.

Footprint

The area on earth illuminated by a satellite antenna. Its shape and size depend upon a number of factors including the antenna design and the satellite's angle of elevation.

Forward Error Correction (FEC)

A transmission error detection and correction technique which, besides *detecting* errors, also *corrects* errors at the receive end of the link. This is achieved without recourse to re-transmission by the receiver (cf Error Detection and Feedback System).

Frequency Division Multiple Access (FDMA)

The earliest and still the commonest form of multiple access. The transponder bandwidth is divided into several subbands, each centred on its own carrier. A number of transmission channels are then multiplexed onto each carrier. This is in contrast to SCPC where each carrier only handles one channel.

Frequency Division Multiplexing (FDM)

A multiplexing method deriving several channels by slicing a given bandwidth into a number of narrow bandwidth channels (Frequency Slicing).

Frequency Modulation

A widely used low-speed modulation method using different frequencies to represent binary 1 and 0.

Frequency Re-use

Increasing the capacity of a satellite by simultaneous re-use of the whole or of parts of the bandwidth. The principal techniques used are multiple spot beams and beam separation, and polarisation.

Frequency Shift Keying (FSK)	See Frequency Modulation.
Gain	A measure of amplifier power expressed in watts or decibels – more usually the latter.
Gaussian (White) Noise	Background noise present in communications channels due to the electrical disturbance of electrons.
GigaHertz	A GigaHertz equals 10^9 Hertz or 1,000 MegaHertz.
Half-Duplex Channel	A channel which is only capable of transmitting in one direction at a time.
Half-Duplex Transmission	Transmission in both directions but not simultaneously.
HDLC	A link level protocol using a frame and bit structure as opposed to character protocols.
Hertz (Hz)	The international standard measurement of frequency (originally expressed in cycles per second).
Impulsive (Black) Noise	Peaks of noise (usually of short duration and high amplitude) which corrupt the data signals.
Intermodulation	The process whereby adjacent Frequency Division Multiplexed carriers modulate one another. Intermodulation products impair satellite performance.
Isochronous	Transmission where all signals are of equal duration and are sent in a continuous sequence.
Isotropic	When radiation from a source has equal strength in all directions, it is described as isotropic. An omnidirectional antenna generates isotropic radiation.

Modem

In data transmission on analogue circuits, refers to the device which effects the conversion from digital to analogue (modulation), and analogue to digital (demodulation).

Modulation

The process whereby a signal is impressed upon a higher frequency carrier. Other functions such as digital to analogue conversion may be carried out at the same time.

Multi-level Signalling

Instead of using just two values of the chosen carrier characteristic to represent '0', '1' respectively (see Frequency Modulation), several levels are employed. This enables more than one bit to be represented by each signal level, and higher transmission rates to be achieved.

Multiplexing

The process of combining separate signal channels into one composite stream (see Frequency Division Multiplexing and Time Division Multiplexing).

Multipoint Circuit

A network configuration with a central site and several outstations connected by a common serving section (American – Multidrop).

Omnidirectional

Refers to the directional radiation capability of an antenna. An omnidirectional antenna is one which is designed to radiate in all directions.

Orbit Types

(a) Geosynchronous.
     Those orbits which a satellite can traverse in a time equal to the earth's period of revolution.
(b) Geostationary.
     The unique orbit which is circular and a distance of 22,500 miles above

the equator. Satellites in this orbit, and travelling in the same direction as the earth's rotation, complete the orbit in 24 hours and always appear stationary relative to points on earth.

(c) Parking.
An orbit in which the satellite is temporarily placed following launch, in preparation to some further manoeuvre. Commonly a low earth orbit, and employed on shuttle launches.

(d) Transfer.
The elliptical orbit with the earth at one focus, with the apogee distance approximating to the geostationary altitude. Successive firings of the apogee motor convert this to the circular geostationary orbit.

Parity

A means of error detection, usually by adding an additional parity bit to a character or block so that the sum of the binary 1s is either odd or even. Conventionally, ODD parity is a feature of synchronous systems and EVEN parity a feature of asynchronous systems.

Perigee

The point on an elliptical orbit where the satellite is closest to the earth.

Phase Modulation

A modulation technique in which the phase angle of the signal is the characteristic which is varied to represent the data being sent. Often combined with amplitude to effect multi-level signalling.

Point-to-Point Link

The simplest physical network configuration, comprising a communication path between two points only.

Protocols

The set of rigorously specified rules which are necessary to ensure the disciplined and accurate transmission and exchange of information across a communication channel.

Pulse Code Modulation (PCM)

A method of transmitting analogue speech in a digital form over a transmission link. The speech bandwidth is converted into a 64 Kbits/sec bit stream.

Quantising Noise/Error

A distortion which arises in PCM due to the difference between the quantisation levels used and the actual signal being sampled.

Residual Error Rate

The undetected error remaining after error detection/correction processes have been carried out.

Serial Transmission

The sequential transmission, on a bit-by-bit basis, of the bits making up a character.

Sideband

The resultant upper and lower frequency bands around a carrier frequency produced in modulation.

Simplex Transmission

Transmission in one direction only (no return path).

Single Channel/Carrier (SCPC)

A multiple access technique in which each distinct user channel is allocated a specific carrier, instead of a number of channels being multiplexed onto a single carrier (cf FDMA).

Space Segment

The portion of a satellite communications system comprising the satellite and the space transmission links.

Tandem

Network configuration where two or more point-to-point circuits are linked together with transmission effected on an end-to-end basis over all links.

Time Division Multiple Access

A multiple access technique which enables a number of earth stations to share a satellite's capacity by allocating each earth station a fixed time slot, during which it can utilise the whole of the capacity (usually a full transponder).

Time Division Multiplexing (TDM)

A multiplexing method deriving several channels by allocating the transmission link for a limited time (time slicing) to each channel. Statistical or intelligent TDM devices further improve the number of channels obtainable.

Transparent

Originally referred to a system which imposes no restrictions on the code or bit pattern used. Increasingly it is being extended to encompass transparency at higher functional levels.

Transponder

In radio communications, a device which receives a signal, amplifies it, and retransmits it. Many also change the frequency as well. In satellite communications it is applied to the individual transponders on board the satellite.

Appendix C

List of Acronyms

ACK	Acknowledgement (positive)
ADCCP	Advanced Data Communications Control Procedure
ADM	Adaptive Delta Modulation
ADPCM	Adaptive Differential Pulse Code Modulation
ALASCOM	Alaskan Communications Satellite Corporation
ANSI	American National Standards Institute
APCM	Adaptive Pulse Code Modulation
ARABSAT	Arabian Satellite Consortium
ARQ	Automatic Retransmission on Request
ASCII	American Standard Code for Information Interchange
AT&T	American Telephone and Telegraph Corporation
BCC	Block Check Character
BER	Bit Error Rate
BSI	British Standards Institution
BSS	Broadcast Satellite Service
BT	British Telecom
BTI	British Telecom International
CATV	Cable Television
CCIR	International Radio Consultative Committee
CCITT	The International Telegraph and Telephone Consultative Committee
CEPT	European Conference of Postal and Telecommunications Administrations
CERN	Centre for European Nuclear Research
COMSAT	Communications Satellite Corporation

DAMA	Demand Assignment Multiple Access
dB	Decibel
DBS	Direct Broadcast Service
DCE	Data Circuit Terminating Equipment
DES	Data Encryption Standard
DM	Delta Modulation
DPCM	Differential Pulse Code Modulation
DSI	Digital Speech Interpolation
DTE	Data Terminal Equipment

EIRP	Equivalent Isotropic Radiated Power
ELV	Expendable Launch Vehicle
ENT	Equivalent Noise Temperature
ESA	European Space Agency
EUTELSAT	European Telecommunications Satellite Organisation

FAX	Facsimile
FCC	Federal Communications Commission (USA)
FDM	Frequency Division Multiplexing
FDMA	Frequency Division Multiple Access
FEC	Forward Error Correction
FSK	Frequency Shift Keying
FSS	Fixed Satellite Service

| G | Gain |
| GHz | GigaHertz |

| HDLC | High-Level Data Link Control |
| Hz | Hertz (cycles per second) |

IF	Intermediate Frequency
IFRB	International Frequency Registration Board
INMARSAT	International Maritime Satellite Organisation
INTELSAT	International Telecommunications Satellite Consortium
ISDN	Integrated Services Digital Network
ISO	International Organisation for Standardisation
ITU	International Telecommunication Union

| KHz | KiloHertz |

MATV	Master Antenna TV
MHz	MegaHertz
MODEM	Modulator/Demodulator
MSS	Mobile Satellite Service
NAK	Negative Acknowledgement
NASA	National Aeronautics and Space Administration
NORAD	North American Aerospace Defence Command
NTU	Network Terminating Unit
PABX	Private Automatic Branch Exchange
PAX	Private Automatic Exchange
PBX	Private Branch Exchange
PCM	Pulse Code Modulation
PSK	Phase Shift Keying
PSS	UK Public Packet Switched Service
PSTN	Public Switched Telephone Network
PTT	Postal, Telegraph and Telephone Authority
QPSK	Quaternary Phase Shift Keying
RCA	Radio Corporation of America
RF	Radio Frequency
SCPC	Single Channel Per Carrier
SDLC	Synchronous Data Link Control
SMATV	Satellite Master Antenna TV
SPADE	Single-channel-per-carrier PCM multiple-Access Demand assignment Equipment
SS/TDMA	Satellite Switched TDMA
STS	Satellite Transport Service
TASI	Time Assigned Speech Interpolation
TDM	Time Division Multiplexing
TDMA	Time Division Multiple Access
TELESAT	Telecommunications Satellite Corporation
TTC	Tracking, Telemetry and Command Subsystem
TVRO	Television Receive Only
TWT	Travelling Wave Tube
WARC	World Administrative Radio Conference

Index

analogue-to-digital conversion 185-187
analogue transmission 174-176
antennae
— beam shaping 109-112
— EIRP 101
— footprint 109-112
— gain 100-101
— general principles 98-100
— performance 100-101
antennae, types of 101-109
— global/earth coverage 102
— hemi/zone 103-104
— omnidirectional 98, 102
— spot beams 104-106
— steerable 106
APOLLO project 293
applications 279-296
— broadcast TV 283-285
— classical telephony 280-282
— commercial and allied applications 286-296
— data transmission and
    file transfer 287-290
— direct broadcast TV 285
— document delivery and
    distribution 292-293
— information distribution 295
— mobile communications 285-286

— remote printing and publishing   290-292
— tele-education   295-296
— video conferencing   294-295

baseband processing   184-192
— analogue-to-digital conversion   185-187
— digital speech compression   187
— Digital Speech Interpolation (DSI) 188
— encryption   190-192
— explanation   184-185
— Forward Error Correction (FEC)   189-190
— video compression   187-188
basic principles   24-34
Bell Laboratories   23
— and TELSTAR   44
Bit Error Rate (BER)   162-163
business applications   286-296

Clarke, Arthur C   23
COMSAT
— protocol investigations   262
— role   45, 214, 242

delay
— data   248-251
— speech   238-240
Demand Assignment Multiple Access
  (DAMA)
— basic principles   33, 228
— with SPADE   229-230
— with TDMA   230-231
digital speech compression   187
Digital Speech Interpolation (DSI)   188
digital transmission   174-176
— on satellite links   175
direct broadcasting by satellite
— concept   46
— impact   56-57

earth segment   28, 30-34, 192-199
  – interfacing and transmission
      standards   196
  — shared versus private
      earth stations   193-194
  — terrestrial access arrangements   192-199
  — user interface   194-195
earth stations
  — classified   202-203
  — functional description   203-208
  — Goonhilly   26, 31
  — Madley   26
  — performance   203
  — standards   203
echo, speech   240-242
encryption   190-192
Equivalent Isotropic Radiated Power
  (EIRP)   101
evolution of satellite communications   41-65
  — direct broadcast television   56-57
  — expansion and growth   43-46
  — experimental phase   41
  — first geostationary satellite   43
  — impact of broadcast television   53-54
  — INTELSAT and the INTELSAT
      satellites   47-50
  — international services   46-47
  — mobile services   54-55
  — national and regional services   50-53
  — small-dish systems   55-56
  — specialised business services   57-59
  — technological trends   59-61

Forward Error Correction (FEC)   189-190, 262
frequency
  — allocation and co-ordination   141-145
  — assignment   135-139
  — choice of   145-147
  — interference   147-148

— plan   139
Frequency Division Multiple Access
  (FDMA)   33, 162, 210-214
frequency hopping   232-233
frequency re-use   113

geostationary orbit   24, 71
geosynchronous orbit   71
GigaHertz   26
Goddard, Robert H   24
Goonhilly earth station complex   26, 31

Hamming, C   189

INMARSAT   45
— role   54
INTELSAT
— establishment and role   46-47
— satellites   47-50
— statistics   47
— UK gateway earth station   198-200
interference   147-148
intermodulation   213
ITU
— organisation and role   142-145
— service classification   145

Keppler, J   68

launch procedures   89-93
— expendable rocket launch   90
— measurement and monitoring phase 91
— shuttle launch   89-90
launch vehicles and rockets
— ARIANE   83
— ATLAS/CENTAUR,
    THOR/DELTA   83, 85
— Space Shuttle   83-85
— TITAN/CENTAUR   86-88

link budget   157-160

Madley earth station complex   26
maritime communications   54
Mercury Communications   298, 314
mobile satellite communications
— evolution   54
— INMARSAT system   54
modulation   167-174
— amplitude   169
— frequency   169
— on satellite links   174
— other techniques   170-174
— phase   169
— pulse code (PCM)   169-170
— quaternary phase shift keying
     (QPSK)   172
multiple access techniques   208-226
— classification   209-210
— compared   224-226
— frequency division (FDMA)   33, 162, 210-214
— frequency hopping   232-233
— random access   231
— single channel per carrier
     (SCPC)   162, 214-216
— time division (TDMA)   33, 94, 162, 216-223
multiplexing   177-183
— frequency division   179
— hierarchical schemes   180
— on satellite links   181-183
— space division   177
— time division   179-180

Newton, Sir Isaac   68
noise
— combatting   156-157
— measurement   153
— signal-to-noise ratio   153
— sources   154-156

orbits
— angle of elevation 75
— drift and station keeping 93-94
— eclipses 80-82
— geosynchronous, geostationary 71-72
— Kepler's Laws 68
— Newton's Laws 68
— orbital dynamics 67-68
— orbital parameters 68-69
— types 68-69
organisations
— ALASCOM 45, 51
— Arianspace 86
— AT&T 44, 303
— Bell Laboratories 23, 44, 188
— British Aerospace 53
— British Telecom 298
— British Telecom International 298
— CCIR 143
— CCITT 143, 196
— CERN 289
— COMSAT 44-45, 214, 244, 262
— ESA 86
— EUTELSAT 52, 282, 287, 293
— FCC 44
— IFRB 143
— INMARSAT 45, 54
— INTELSAT 46-50, 75, 102, 104, 106,
110, 193, 198, 202, 203,
214, 273-274, 281-282,
300-302
— ITU 74, 142-145
— Mercury Communications 298
— NASA 105
— NORAD 131
— RCA 51
— SBS 59, 196, 244, 262, 294-295
origins 23-27

Pierce, J R   23
power supply system   118-121
— solar arrays   119-121
— solar energy   118-119
protocol implementation
— conclusions, recommendations   264-265
— considerations   262-264
protocols
— and satellite transmission   246-248
— binary synchronous,
    basic mode   253
— compared   251-262
— continuous ARQ   253-262
— delay implications   248-251
— functions   243-245
— information feedback
    (echoplexing)   251
— purpose   243

regulations
— Europe   272
— integration of terrestrial and
    satellite services   274
— pressures for de-regulation   271
— space segment access
    constraints   273
— space segment competitive
    pressures   273-274
— tariff implications   272
— USA   44-45, 272
reliability   125-126
role of satellite communications
— competition from terrestrial
    services   268
— complementary role   274-278
— geographical factors   268
— interconnection potential   278
— political considerations   270

— regulatory environment   271-274
— role of television   270
— social and economic factors   269
— supply flexibility   278

satellite broadcast television   53-54, 283-285
Satellite Business Systems (SBS)   59, 244, 264
satellite communications
— basic principles   24-34
— broadcast   34
— delay and echo   38-39, 237-238
— distance insensitivity   37
— evolution   41-65
— geographical flexibility   35
— origins   23-27
— transmission capacity   37
— unique properties   34-39
satellite construction
— antennae subsystem   95-113
— power supply system   118-121
— satellite life-span   126-127
— space debris   131
— space environment   127-130
— spacecraft reliability   125-126
— stabilisation and attitude
    control   121-123
— tracking, telemetry and
    command   124
— transponder subsystem   113-118
satellite launching and positioning
— choice of launch vehicles   88
— expendable launch vehicles   85-88
— the Space Shuttle   83-84
satellite life-span   126-127
satellite services
— direct broadcast television   46, 56-57
— European business services   200, 216
— in Alaska   51
— in Europe   52-53
— INTELSAT business services   200, 203

— international telecommunications   280-282
— ITU classification   145
— mobile and maritime
    communications   54
— personalised services   317-319
— regional and national
    telecommunications   50-53
— satellite broadcast television   53-54
— specialised business services   57-59
— UK service plans   298-304
satellite spacing   73-75
satellite switched TDMA (SS/TDMA)   233
satellites
— ANIK   51
— ARABSAT   53
— ATS series   51, 105
— COMSTAR   51
— COURIER   42
— Early Bird   47-48
— ECS-1, ECS series   52, 109, 139, 190-191, 200
— EXPLORER 1   41
— INTELSAT   47-50
— LANDSAT   288
— L-SAT, OLYMPUS   53
— MARECS   54
— MOLNYA   43
— OTS   52
— PALAPA   53
— RELAY   42
— SATCOM   51
— SCORE   41
— SPUTNIK   41
— SYNCOM   43
— TELECOM-1   287, 299-300
— TELSTAR   42, 44
— UNISAT   109, 112, 135
— WESTAR   51
SatStream services   298-299
— economics   304-316
— facilities   303-304

— presentation   302-303
— tariffs   309
Shannon, C E   160
signal-to-noise ratio   30, 157-161
Single Channel Per Carrier (SCPC)   162, 214
solar cells   118-121
space environment   127-131
— effect on materials,
    lubrication   128-129
— heat transfer considerations   129
— solar and other galactic
    radiation sources   129-130
— space debris   130-131
space segment   28
— antennae   26
— satellite functional
    components   28-30
— transmission link   26
space segment access methods   208-224
— classification   209-210
— FDMA   33, 162, 210-214
— SCPC   162, 214
— SPADE   214-216
— TDMA   33, 94, 162, 216-224
Space Shuttle   83-84
SPADE   214-216, 230-231
stabilisation and attitude control   121-123
— position measurement
    and accuracy   122-123
— types of stabilisation   121-122
STELLA project   289

technological trends
— on-board switching   136, 323
— satellite switched TDMA   233
— satellite technology   59-61
terrestrial access arrangements
— ECS-2 business services   200
— UK INTELSAT gateway   200

thrust subsystem   125
Time Assigned Speech Interpolation
   (TASI)   188
Time Division Multiple Access (TDMA)   33, 94, 162, 216-224
Tracking, Telemetry and Command (TTC)   102, 124-125
transmission capacity
  — examples   37, 161-162
  — of the ECS-2 business
     transponders   162
  — satellite and optical fibre
     compared   37
  — theoretical capacity   160-161
transmission losses   149-152
  — atmospheric   150-152
  — free space   149-150
transmission performance   157-163
  — channel bit error rate   162-163
  — link budget   157-160
  — overall system performance   163-165
transmission principles   167-183
  — analogue and digital   174-176
  — modulation and modulation
     techniques   167-174
  — multiplexing   177-183
transponder   26, 113-118
transponder subsystem   113-118
  — functional components   115-118
  — high power amplifiers   117-118
  — travelling wave tube   117-118
travelling wave tube   117-118
Tsiolkovsky, Konstantin   23

UK business services   298-302
  — SatStream   298
UK INTELSAT access   198-200
UK satellite service economics   304-316
  — tariffs   310
  — UK mainland   308-313
  — US example   305

— within continental Europe   313-314
UNIVERSE project   262

video compression   187-188
Viterbi, A J   190
Viterbi Convolutional Encoder   190
von Braun, Wernher   24

wartime experience   24

Mrs Pepperpot to the Rescue

and other stories

MRS PEPPERPOT
TO THE RESCUE

and other stories by
ALF PRØYSEN

Translated by Marianne Helweg
Illustrated by Björn Berg

HUTCHINSON JUNIOR BOOKS

Hutchinson Junior Books Ltd
3 Fitzroy Square, London W1

An imprint of the Hutchinson Publishing Group

London Melbourne Sydney Auckland
Wellington Johannesburg and agencies
throughout the world

First published May 1963
Second impression March 1964
Third impression April 1966
Fourth impression March 1967
Fifth impression February 1969
Sixth impression March 1971
Seventh impression February 1972
Eighth impression September 1974
Ninth impression June 1977

Printed in Great Britain by litho at
The Anchor Press Ltd and bound by
Wm Brendon & Son Ltd
both of Tiptree, Essex

ISBN 0 09 067920 2

Contents

Mrs Pepperpot to the Rescue, 9

Mrs Pepperpot on the Warpath, 24

The Nature Lesson, 35

The Shoemaker's Doll, 47

Mrs Pepperpot is taken for a Witch, 55

The Little Mouse who was Very Clever, 67

Mrs Pepperpot's Birthday, 71

The Dancing Bees, 77

How the King Learned to Eat Porridge, 88

Mrs Pepperpot Turns Fortune-Teller, 96

The Fairy-Tale Boy, 107

The Ski-Race, 116

Mrs Pepperpot to the Rescue

and other stories

Mrs Pepperpot to the Rescue

WHEN IT IS breaking-up day at the village school, and the summer holidays are about to begin, all the children bring flowers to decorate the school. They pick them in their own gardens or they get them from their uncles and aunts, and then they carry their big bunches along the road, while they sing and shout because it is the end of term. Their mothers and fathers wave to them from the windows and wish them a happy breaking-up day.

But in one window stands a little old woman who just watches the children go by. That is Mrs Pepperpot.

She has no one now to wish a happy breaking-up day, for all her own children are long since grown up and gone away, and none of the young ones think of asking her for flowers.

At least, that's not quite true; I do know of one little girl who picked flowers in Mrs Pepperpot's garden. But that was several years ago, not long after the little old woman first started shrinking to the size of a pepperpot at the most inconvenient moments.

9

That particular summer Mrs Pepperpot's garden was fairly bursting with flowers: there were white lilac with boughs almost laden to the ground, blue and red anemones on strong, straight stalks, poppies with graceful nodding yellow heads and many other lovely flowers. But no one had asked Mrs Pepperpot for any of them, so she just stood in her window and watched as the children went by, singing and shouting, on their way to the breaking-up day at school.

The very last to cross the yard in front of her house was a little girl, and she was walking, oh, so slowly, and

carried nothing in her hands. Mrs Pepperpot's cat was lying on the doorstep and greeted her with a 'Miaow!' But the little girl only made a face and said, 'Stupid animal!' And when Mrs Pepperpot's dog, which was chained to the wall, started barking and wagging his tail the little girl snapped, 'Hold your tongue!'

Then Mrs Pepperpot opened the window to throw a bone out to the dog and the little girl whirled round and shouted angrily, 'Don't you throw that dirty bone on my dress!'

That was enough. Mrs Pepperpot put her hands on her hips and told the little girl that no one had any right to cross the yard in front of her house and throw insulting words at her or her cat and dog, which were doing no harm to anybody.

The little girl began to cry. 'I want to go home,' she sobbed. 'I've an awful pain in my tummy and I don't want to go to the breaking-up party! Why should I go when I have a pain in my tummy?'

'Where's your mother, child?' asked Mrs Pepperpot.

'None of your business!' snapped the girl.

'Well, where's your father, then?' asked Mrs Pepperpot.

'Never you mind!' said the girl, still more rudely. 'But if you want to know why I don't want to go to school today it's because I haven't any flowers. We haven't a garden, anyway, as we've only been here since Christmas. But Dad's going to build us a house now that he's working at the ironworks, and then we'll have a garden. My mum makes paper flowers and does the paper round, see? Anything more you'd like to know? Oh well, I might as well go to school, I suppose. Teacher can say what she likes—I don't care! If *she'd* been going from school to school for three years she wouldn't know much either! So blow her and her flowers!' And the little girl stared defiantly at Mrs Pepperpot.

Mrs Pepperpot stared back at the little girl and then

she said: 'That's the spirit! But I think I can help you with the flowers. Just you go out in the garden and pick some lilac and anemones and poppies and anything else you like. I'll go and find some paper for you to wrap them in.'

So the girl went into the garden and started picking flowers while Mrs Pepperpot went indoors for some paper. But just as she was coming back to the door she shrank!

Roly Poly! And there she was, tucked up in the paper like jam in a pudding, when the little girl came running back with her arms full of flowers.

'Here we are!' shouted the little girl.

'And here *we* are!' said Mrs Pepperpot as she disentangled herself from the paper. 'Don't be scared; this is something that happens to me from time to time, and I never know when I'm going to shrink. But now I've got an idea; I want you to pop me in your satchel and take me along with you to school. We're going to have a game with them all! What's your name, by the way?'

'It's Rita,' said the little girl who was staring at Mrs Pepperpot with open mouth.

'Well, Rita, don't just stand there. Hurry up and put the paper round those flowers. There's no time to lose!'

When they got to the school the breaking-up party was well under way, and the teacher didn't look particularly pleased even when Rita handed her the lovely bunch of flowers. She just nodded and said, 'Thanks.'

'Take no notice,' said Mrs Pepperpot from Rita's satchel.

'Go to your desk,' said the teacher. Rita sat down with her satchel on her knee.

'We'll start with a little arithmetic,' said the teacher. 'What are seven times seven?'

'Forty-nine!' whispered Mrs Pepperpot from the satchel.

'Forty-nine!' said Rita.

This made the whole class turn round and stare at Rita, for up to now she had hardly been able to count to thirty! But Rita stared back at them and smiled. Then she stole a quick look at her satchel.

'What's that on your lap?' asked the teacher. 'Nobody is allowed to use a crib. Give me your satchel at once!'

So Rita had to carry it up to the teacher's desk where it was hung on a peg.

The teacher went on to the next question: 'If we take fifteen from eighteen what do we get?'

All the children started counting on their fingers, but Rita saw that Mrs Pepperpot was sticking both her arms and one leg out of the satchel.

'Three!' said Rita before the others had had time to answer.

This time nobody suspected her of cheating and Rita beamed all over while Mrs Pepperpot waved to her from between the pages of her exercise book.

'Very strange, I must say,' said the teacher. 'Now we'll have a little history and geography. Which country is it that has a long wall running round it and has the oldest culture in the world?'

Rita was watching the satchel the whole time, and now she saw Mrs Pepperpot's head pop up again. The little old woman had smeared her face with yellow chalk and now she put her fingers in the corners of her eyes and pulled them into narrow slits.

'China!' shouted Rita.

The teacher was quite amazed at this answer, but she had to admit that Rita was right. Then she made an announcement.

'Children,' she said, 'I have decided to award a treat to the one of you who gave the most right answers. Rita gave me all the right answers, so she is the winner, and she will be allowed to serve coffee to the teachers in the staff-room afterwards.'

Rita felt very pleased and proud; she was so used to getting meals ready when she was alone at home that she was sure she could manage this all right. So, when the other children went home, she took her satchel from the teacher's desk and went out into the kitchen. But, oh dear, it wasn't a bit like home! The coffee-pot was far too big and the huge cake with icing on it was very different from the plate of bread-and-dripping she usually got ready for her parents at home. Luckily the cups and saucers and plates and spoons had all been laid out on the table beforehand. All the same, it seemed too much to Rita, and she just sat down and cried. In a moment she heard the sound of scratching from the satchel, and out stepped Mrs Pepperpot.

'If you're the girl I take you for,' said the little old woman, putting her hands on her hips, 'you won't give up half-way like this! Come on, just you lift me up on the table, we'll soon have this job done! As far as I could see from my hiding place, there are nine visiting teachers and your own Miss Snooty. That makes two cups of water and two dessertspoons of coffee per person—which makes twenty cups of water and twenty dessertspoons of coffee in all—right?'

'I think so. Oh, you're wonderful!' said Rita, drying her tears. 'I'll measure out the water and coffee at once, but I don't know how I'm going to cut up that cake!'

'That'll be all right,' said Mrs. Pepperpot. 'As far as I can see the cake is about ninety paces—my paces—round. So if we divide it by ten that'll make each piece nine paces. But that will be too big for each slice, so we'll divide nine by three and make each piece three paces thick. Right?'

'I expect so,' said Rita, who was getting a bit lost.

'But first we must mark a circle in the middle of the cake,' went on Mrs Pepperpot. 'Lift me up on your hand, please.'

Rita lifted her carefully on to her hand.

'Now take me by the legs and turn me upside down. Then, while you swing me round, I can mark a circle with one finger in the icing. Right; let's go!'

So Rita swung Mrs Pepperpot round upside down and the result was a perfect little circle drawn right in the middle of the cake.

'Crumbs are better than no bread!' said Mrs Pepperpot as she stood there, swaying giddily and licking her finger. 'Now I'll walk right round the cake, and at every third step I want you to make a little notch in the icing with the knife. Here we go!

'One, two, three, notch!
One, two, three, notch!
One, two, three, notch!'

And in this way Mrs Pepperpot marched all round the cake, and Rita notched it so that it made exactly thirty slices when it was cut.

When they had finished someone called from the staff-room: 'Where's that clever girl with the coffee? Hurry up and bring it in, dear, then you can fetch the cake afterwards.'

Rita snatched up the big coffee-pot, which was boiling now, and hurried in with it, and Mrs Pepperpot stood listening to the way the teachers praised Rita as she poured the coffee into the cups with a steady hand.

After a while she came out for the cake. Mrs Pepperpot clapped her hands: 'Well done, Rita! There's nothing to worry about now.'

But she shouldn't have said that, for while she was listening to the teachers telling Rita again how clever she was, she suddenly heard that Miss Snooty raising her voice:

'I'm afraid you've forgotten two things, dear,' she said.

'Oh dear!' thought Mrs Pepperpot, 'the cream-jug and the sugar-bowl! I shall have to look and see if they are both filled.'

The cream-jug was full, but when Mrs Pepperpot leaned over the edge of the sugar-bowl she toppled in! And at the same moment Rita rushed in, put the lid on the sugar-bowl and put it and the cream-jug on a little tray. Then she turned round and went back to the staff-room.

First Mrs Pepperpot wondered if she should tell Rita where she was, but she was afraid the child might drop the tray altogether, so instead she buried herself well down in the sugar-bowl and hoped for the best.

Rita started carrying the tray round. But her teacher hadn't finished with her yet. 'I hope you remembered the sugar-tongs,' she said.

Rita didn't know what to say, but Mrs Pepperpot heard the remark, and when the visiting head teacher took the lid off, Mrs Pepperpot popped up like a jack-in-the-box holding a lump of sugar in her outstretched hand. She stared straight in front of her and never moved

an eyelid, so the head teacher didn't notice anything odd. He simply took the sugar lump and waved Rita on with the tray. But his neighbour at the table looked hard at Mrs Pepperpot and said: 'What very curious sugar-tongs—I suppose they're made of plastic. Whatever will they think of next?' Then he asked Rita if she had brought them with her from home, and she said yes, which was strictly true, of course.

After that everyone wanted to have a look at the curious sugar-tongs, till in the end Rita's teacher called her over.

'Let me have a look at those tongs,' she said. She reached out her hand to pick them up, but this was too much for Mrs Pepperpot. In a moment she had the whole tray over and everything fell on the floor. The

cream-jug was smashed and the contents of the sugar-bowl rolled under the cupboard, which was just as well for Mrs Pepperpot!

But the teacher thought it was she who had upset the tray, and suddenly she was sorry she had been so hard on the little girl. She put her arms round Rita and gave her a hug. 'It was all my fault,' she said. 'You've been a very good little parlourmaid.'

Later, when all the guests had gone, and Rita was clearing the table, the teacher pointed to the dark corner by the cupboard and said, 'Who is that standing there?'

And out stepped Mrs Pepperpot as large as life and quite unruffled. 'I've been sent to lend a hand with the washing-up,' she said. 'Give me that tray, Rita. You and I will go out into the kitchen.'

When at last the two of them were walking home, Rita said, 'Why did you help me all day when I was so horrid to you this morning?'

'Well,' said Mrs Pepperpot, 'perhaps it was because you *were* so horrid. Next time maybe I'll help that Miss Snooty of yours. She looks pretty horrid too, but she might be nice underneath.'

Mrs Pepperpot on the Warpath

It was the day after Mrs Pepperpot had helped Rita at the school party, and the little old woman was in a terrible rage. You see, if there's anything Mrs Pepperpot hates, it's people being unkind to children. All night she had been thinking about it, and now she had made up her mind to go and tell Rita's teacher just what she thought of her. So she put on her best hat and her best frock, straightened her back and marched off to the school.

'I hope I don't shrink this time,' she thought, 'but it's not likely to happen two days running. Anyway, today I must have my say or I shall burst. Somebody's going to say she's sorry or my name's not Pepperpot!'

She had reached the school gate and swung it open. Then she walked up to the teacher's front door and knocked twice smartly. Then she waited.

No one said, 'Come in!'

Mrs Pepperpot knocked again, but there was still no answer. So she decided to try the latch. 'If the door isn't locked I shall go straight in,' she said to herself. She pressed the latch and the door opened. But no sooner had she put a foot over the threshold than she shrank and fell head over heels into a travelling-rug which was rolled up on the floor just inside the door! Next to it stood a suitcase and a hatbox.

'Oh, calamity!' cried Mrs Pepperpot, 'let's hope she's not in after all now!' But she was unlucky, for now she could hear footsteps in the corridor and the teacher came towards the front door dressed in her going-out clothes.

'What an old dolt I am!' thought Mrs Pepperpot. 'Fancy me not remembering the summer holidays have started today and she'll be going away, of course. Oh well, she's not gone yet. If I can manage to stay near her

for a little while longer I may still get my chance to give her a piece of my mind.' So she hid in the rug.

The teacher picked up the suitcase in one hand, then she threw the travelling-rug over her shoulder and picked up the hatbox in the other hand and walked out of the house, closing the door behind her. And Mrs Pepperpot? She was clinging for dear life to the fringe of the rug and she was still as angry as ever.

'Very nice, I must say!' she muttered. 'Going away on a holiday like this without a thought for Rita and all the harm you did her. But you wait, my fine lady, very soon it'll be my turn to teach you a thing or two!'

26

The teacher walked briskly on, with Mrs Pepperpot dangling behind her, till they got to the station. Then she walked over to the fruit-stall and put the rug down on the counter, and Mrs Pepperpot was able to slip out of it and hide behind a bunch of flowers.

The teacher asked for two pounds of apples.

'That's right!' fumed Mrs Pepperpot to herself. 'Buy two pounds of apples to gorge yourself with on the train!'

'And eight oranges, please,' continued the teacher.

'Worse and worse!' muttered Mrs Pepperpot.

'And three pounds of bananas, please,' said the teacher.

Mrs Pepperpot could hardly contain herself: 'If I was my proper size now, I'd give you apples and oranges and bananas, and no mistake!'

Then the teacher said to the lady in the fruit-stall:

'Do you think you could do me a favour? I want all this fruit to go to one of my pupils, but I have to catch the train, so I've no time to take it to her myself. Could you deliver it to Rita Johansen in the little house by the church and tell her it's from me?'

On hearing this, Mrs Pepperpot's ears nearly fell off with astonishment. It was just as if someone had taken a sweet out of her mouth and left her nothing to suck; what was she going to say now?

'I'll do that for you with pleasure, miss,' said the fruit-lady. 'That'll be twelve shillings exactly.'

'Oh dear!' exclaimed the teacher, rummaging in her purse, 'I see I shan't have enough money left after buying my ticket. Would you mind if I owed you the twelve shillings till I come back from my holidays?'

'The very idea! Asking me to deliver goods you can't

even pay for! I shall have to have the fruit back, please,' the fruit-lady said, and held out her hand.

The teacher said she was sorry, put the bag of fruit back on the counter and went off to board her train, but Mrs Pepperpot had taken the chance to jump into the bag.

Silently she wished the teacher a good holiday: 'You're not so bad after all, and you needn't worry; I'll see that Rita gets her bag of fruit somehow. But *somebody's* going to get the edge of my tongue before the day's out!'

Of course the fruit-lady could no more hear what Mrs Pepperpot was saying than the teacher could. She was busy getting ready to shut up shop and go home. But when she had put her hat on and opened the door she suddenly turned round and picked up the bag of fruit on the counter.

Mrs Pepperpot had just been wondering if she was going to be locked in the fruit-stall all night, and now here she was, being taken on another journey!

'I suppose you're going to eat all this yourself, you selfish old thing, you!' thought Mrs Pepperpot, getting worked up again. 'The teacher may be snooty, but at least she has a kind heart underneath. You're just plain mean! But just you wait till I grow again!'

The fruit-lady walked on and on, until at last Mrs

29

Pepperpot could hear her opening a door and going into a room. There she set the bag down with a thump on the table, and Mrs Pepperpot was able to climb over an orange and peep out of the top.

She saw a man banging on the table, and he was as cross as a sore bear. 'What sort of time is this to come home?' he roared. 'I've been waiting and waiting for my supper. Hurry up now! What's in that bag, anyway?'

'Oh, it's only some fruit for a little girl,' said his wife. 'The school-teacher wanted to send it to Rita Johansen, but she found she hadn't enough money, so I took it back. Then when she'd gone I felt sorry, so I thought I'd take it along to the child myself.'

30

This time Mrs Pepperpot was really amazed: 'Well, I never!' she gasped, 'here's another one who turns out to be nice. Still, I'm sure her husband won't; he looks as if he could do with a good ticking off!'

The fruit-lady's husband certainly was a cross-patch and no mistake. He banged his fist on the table and shouted that no wife of his was going to waste money and time running errands for silly school-teachers and brats.

'Give me that bag!' he roared. 'I'll take it right back to the shop this minute!' And he snatched up the bag from the table. Poor Mrs Pepperpot was given an awful shaking and landed up jammed between two bananas.

Taking long strides, the man walked off down the road.

'Bye-bye, fruit-lady!' whispered Mrs Pepperpot. 'You have a nasty husband, but I'll deal with him shortly, don't you worry!'

Squeezed and bruised, the little old woman lay there in the bag while the man strode on. But after a while he walked more slowly and at last he stopped at a house and knocked on the door.

'Surely this isn't the station?' wondered Mrs Pepperpot.

She heard the door open and the man spoke: 'Are you Rita Johansen?'

Then she heard a little girl's voice, 'Yes, that's me.'

'Your teacher sent you this,' said the man and handed over the bag; 'it's fruit.'

'Oh, thank you!' said Rita. 'I'll just go and get a bowl to put it in.' And she set the bag on a chair.

'That's all right,' said the man, and he turned on his heel and walked away.

When Rita came back with the bowl she thought she heard the door close, but she didn't take much notice in her eagerness to see what the teacher had sent her.

But it was actually Mrs Pepperpot who had slipped out, for she was now her usual size and she wanted time to think; it had all been so surprising and not at all what she expected. As she walked she began to hurry. For now she knew who was going to get a piece of her mind, and rightly so! Someone who made her more angry than anyone else just now!

When she got home she marched straight to the mirror. Putting her hands on her hips she glared at the little old woman she saw there. 'Well!' she said, 'and who do you think you are, running round the country-side, poking your nose in where you're not wanted? Is it any of your business, may I ask, who the school-teacher buys fruit for? What d'you mean by hiding in people's travelling-rugs and spying on them? You ought to be ashamed of yourself, an old woman like you, behaving like a senseless child. As for the fruit-lady, why shouldn't she be cross? How was she to know if she could trust the teacher? And her husband; I suppose he can bang his fist on his own table if he likes without you interfering? Are you listening? Wouldn't you be pretty mad if you'd come home hungry and the wife wasn't there to cook your .

meal, eh? I'm disgusted with you! *They* were sorry for what they did and made amends, all three of them, but *you*, you just stand there glaring at me as if nothing had happened. Wouldn't it be an idea to say you were sorry?'

Mrs Pepperpot turned her back on the mirror and took a deep breath. 'That's better!' she said. 'I've got it all off my chest at last. Now I can give my tongue a rest and get on with the housework.'

But first she took one more look in the mirror, smiled shyly and bobbed a little curtsy.

'I'm sorry!' she said.

And the little old woman in the mirror smiled back at her and bobbed a little curtsy too.

The Nature Lesson

EVERY MORNING, when Mrs Pepperpot sits at her window with her after-breakfast cup of coffee, she watches a little boy who always walks across her yard on his way to school. The boy's name is Olly and he and Mrs Pepperpot are very good friends, though not in the way grown-ups usually are friends with children. Quite often Olly rushes past Mrs Pepperpot's window without even saying 'Good morning', because he is in such a hurry. But then Mrs Pepperpot has never even asked him his name or how old he is or what he wants for Christmas. She just watches him every morning and says to herself, 'There goes the little boy on his way to school.' As for Olly, he just glances up at her window and thinks, 'There's the old woman, drinking her coffee.'

Now with animals it is different: if Olly sees the cat sitting on Mrs Pepperpot's door-step he can't resist stopping to stroke her. He'll even sit down on the door-step and talk to her.

'Hullo, pussy,' he'll say. 'There's a lovely pussy!' And then, of course, he has to go and see the dog outside his kennel as well, in case he should get jealous.

'Hullo, boy! Good dog, good dog! You didn't think I'd forgotten you, did you? Oh, I wouldn't do that! There's a good dog!' And by the time he's made a fuss of them both he's late for school.

This is Olly's trouble: he's *very* fond of animals. He loves to play hide-and-seek with the squirrel he sees on his way to school, or to have a whistling-match with a blackbird. And as for *rainy* days, well, he spends so much time trying not to step on the worms wriggling by the puddles in the road that he's *always* late for school.

This won't do, of course, and when he's late his teacher gets cross, and she'll say, 'It's all very well being fond of animals; it's quite right that you should be, but it's no excuse for being late for school.'

But that wasn't what I was going to tell you. What I was going to tell you was how Mrs Pepperpot had a nature lesson one day. So here we go!

It was a lovely spring day, and Mrs Pepperpot was sitting by the window as usual, enjoying her cup of coffee and watching Olly come across the yard. He was walking rather briskly this time—watching some bird or animal had probably made him late again—so he had

only time to say 'Hullo, puss!' to the cat on the door-step and 'Hi, boy!' to the dog by the kennel.

But suddenly he stopped dead, turned round on his heel and started running back across the yard. Mrs Pepperpot had just come to the door to give the dog his breakfast and Olly rushed past her as fast as he could go.

Mrs Pepperpot called to him: 'Whatever's the matter with you, boy? The police after you?'

'Forgot my nature textbook!' answered Olly over his shoulder, and started off again.

'Wait a minute!' called Mrs Pepperpot. Olly stopped. 'You can't go all the way home again now; you'll be much too late for school. No, you go on and *I'll* go back for your book and bring it to you at school.'

Olly shuffled his feet a bit and looked unhappy; he didn't much like the idea of an old woman turning up in school with his nature textbook.

'Don't stand there shuffling, boy!' said Mrs Pepperpot. 'Where did you leave the book?'

'On the window-sill,' he answered; 'the window is open.'

'All right. Where do you want me to put the book when I get to the school? Come on, hurry up; we haven't got all day!' said Mrs Pepperpot, trying to look severe.

'There's a hole in the wall, just by the big birch tree; there's an old bird-nest there you can put it in.'

'In the old nest in a hole in the wall by the birch tree; right!' said Mrs Pepperpot. 'Now, off you go and see if you can be on time for a change! I'll see to the rest.'

'Righto!' said Olly and was off before you could say Jack Robinson.

Mrs Pepperpot took off her apron, smoothed her hair and stepped out into the yard. And then, of course, the inevitable thing happened; she shrank!

'This is bad,' thought Mrs Pepperpot, as she peeped over the wet grass by the door-step, 'but I've known worse.' She called to the cat: 'Come here, puss! You'll have to be my horse once again and help me fetch Olly's nature book from his house.'

'Miaow! All right,' said the cat, as she allowed Mrs Pepperpot to climb on her back. 'What sort of a thing is a nature book?'

'It's a book the children use in school to learn about animals,' answered Mrs Pepperpot, 'and one thing it says about cats is that you are "carnivores".'

'What does that mean?' asked the cat.

'That you eat meat, but never mind that now; all you have to do is to take me straight down the road till we get to the stream. Then we take a short cut across. . . .'

But, as they came near the stream, the cat said, 'Doesn't the book say anything about cats not liking to get their feet wet?' And then she stopped so abruptly that poor Mrs Pepperpot toppled right over her head and fell plump into the water!

'Good job I can swim,' spluttered Mrs Pepperpot as she came to the surface, 'humans aren't meant to live under water on account of the way they breathe with their lungs. Phew! It's hard work all the same; I'll take a rest on this stone and see if something turns up.'

While she was getting her breath a tiny animal stuck its nose out of the water, and started snarling at her. Now Mrs Pepperpot knew what that was, but you probably wouldn't, because it only lives in the faraway places, and it is called a lemming. Its fur is dappled brown and fawn, so that it looks a bit like a guinea-pig in summer, but in winter it turns white as the snow around it.

As I say, Mrs Pepperpot knew all about lemmings, so she snarled back at the little creature, making as horrible a noise as she could. 'I'm not afraid of you!' she said, 'though the book says you're the worst-tempered of all the little rodents and don't give way to a fierce dog or even a grown man. But now you can just stop showing off and help me out of this stream like a good lemming.'

'Well, blow me down!' said the lemming. 'I never saw a woman as small as you and with such a loud voice. Get on my back and I'll take you across. Where are you going, by the way?'

'To fetch a nature book from the house over there for a little boy at school,' said Mrs Pepperpot. 'And in that book there is quite a bit about you.'

'Oh? And what does it say?' asked the lemming, crawling out on to the grass with Mrs Pepperpot.

'It says that once every so many years lemmings come down from the mountains in great swarms and eat up all the green stuff they can find till they get to the sea.' Then she stopped, because she remembered that when the lemmings reach the sea in their search for food, thousands of them get drowned.

'We do get rather hungry,' said the lemming; 'as a matter of fact, I'm on my way now to join my mates in a little food-hunt. . . .'

'Couldn't you just take me down to the house?' pleaded Mrs Pepperpot; she didn't like the idea that he might drown in the sea. But the lemming's empty tummy was telling him to go, so he told Mrs Pepperpot she would have to manage by herself, and he ran off muttering to himself about juicy green leaves.

Before Mrs Pepperpot had had time to wonder what

would happen to him, another head appeared above a little wall. This time it was a stoat.

'Hullo, Mr Nosey Parker,' she greeted him, 'what are you looking for?'

'I thought you were a mouse, but I see you're a little old woman, and I don't eat women,' said the stoat. 'Have you by any chance got a silver spoon?' he added.

'I have something you like even better than silver spoons,' answered Mrs Pepperpot, 'a whole packet of tin-tacks, and you can have them if you'll take me to that house over there. I have to fetch a book from the window-sill for a little boy in the school.'

'All right,' said the stoat, 'hop up!'

So Mrs Pepperpot got on his back. But it was a most

uncomfortable journey, because stoats, like weasels, move by rippling their long bodies, and though they have short legs, they can run very fast. Mrs Pepperpot had a job keeping on and was glad when they reached the wall under the window.

The stoat scrambled up to the window-sill, and presently he came back with the book—under his chin.

'Why do you carry the book that way?' asked Mrs Pepperpot.

'How else?' answered the stout. 'I always carry eggs under my chin.'

'Eggs?' Mrs Pepperpot pretended to be surprised. 'I didn't know stoats laid eggs.'

'Ha, ha, very funny!' said the stoat. 'I suppose you don't eat eggs?'

'Oh yes,' said Mrs Pepperpot, 'but I don't steal them out of wild birds' nests.'

'That's my business,' said the stoat. 'Now you'd better think how you're going to get this book to school; I can't carry both you and the book.'

'That's true!' said Mrs Pepperpot. 'I'll have to think of something.'

But it wasn't necessary, for the next moment Mrs Pepperpot was back to her proper size. As she bent down to pick up the book she whispered to the little stoat, 'The tin-tacks will be waiting for you in that nest you robbed in the stone wall by the school.' And she thought she heard him chuckle as he rippled away in the grass.

When she reached the school the bell was ringing for break, and she just had time to pop the book into the empty nest before Olly came running out with the other children. Mrs Pepperpot gave the tiniest nod in the direction of the wall and then she walked briskly away.

But the next morning Olly brought a lovely bone for her dog and from his milk bottle he poured a good saucer-full of milk for her pussy.

Mrs Pepperpot opened the window. 'Would you do something for me this morning?' she asked.

'As long as it won't make me late for school,' answered Olly.

'Good,' said Mrs Pepperpot. Then she fetched a packet of tin-tacks from the toolshed and gave them to Olly. 'Put those in the empty nest in the wall, will you? They're for a friend of mine.'

The Shoemaker's Doll

THERE WAS once a shoemaker who won a doll in a raffle at a bazaar. But a doll was no good to him, living alone as he did in a little cottage where the floor was strewn with old soles and bits of leather and where everything he touched was covered in glue.

For, with all the village shoemaking to be done before he could do his own housework, you can imagine what the shoemaker's home looked like. And now there was this doll. When he had brought it home he stood looking round the little room and wondering where on earth he could put it. At last he decided to set the doll on top of the chest of drawers next to a half-loaf of bread and a rubber boot. Then he went to bed.

In the night he dreamed that the doll came over to his bed and said:

> 'I can scrub, I can sweep,
> Make a bed and house-keep;
> If *you* won't, *I* will!'

When the shoemaker opened his eyes next morning and saw the doll on the chest of drawers he remembered the dream and laughed to himself.

'So you can scrub and sweep and make beds, can you, my little flibberty gibbett? Still, I suppose I had better do it myself to save you spoiling your pretty frock,' he said. So he made his bed, which hadn't been done properly for fifteen years, and underneath he found the old gold watch his grandfather had left him. Until that moment he thought it was lost for ever. After that he washed the floor and found a half-crown lying in a corner behind the cupboard. Then he swept out all the rubbish from the drawer in his work-bench, and what do you think?

Right at the bottom he found a little ring with a red stone in it!

At first the shoemaker was astonished, but then he remembered:

'Yes, yes, of course! It was that little girl from Crag House; she did tell me she had lost her ring here once. She came to ask me about it several times. I'll put it on the shelf here and give it to her when I see her. And now I suppose I'd better tidy up the chest of drawers for you, my little slave-driver!'

He started by removing the rubber boot. But he decided to leave the half-loaf, because he liked to take a bite from time to time while he was mending the shoes.

Later in the day a village woman came in. When she

saw how tidy the room was she clapped her hands and exclaimed: 'Well I never! What a change! You must have had a lot of time to spare getting the place so ship-shape. I suppose you have my shoes ready as well, then?'

The shoemaker pushed his spectacles up on his forehead and stared at the woman. Then he said, 'Come again Tuesday!'

'But it's Tuesday today,' objected the woman.

The shoemaker let his spectacles fall back on his nose.

'Come again Wednesday!' he snapped and went on with his work.

'Oh, you're just the same old lazy sour-puss that you always were!' said the woman, and left him to it.

In the afternoon a little boy started playing in the

street outside the house. He came to the door and asked the shoemaker to button his jacket because his fingers were too cold to manage the buttonhole.

'Come again Thursday!' said the shoemaker in his snappy voice.

'But I'm cold today, mister!' said the little boy.

'Come again Friday!' said the shoemaker.

'You're just a grumpy old toad, so you are!' said the little boy, and went away.

'Good riddance!' said the shoemaker, and he didn't even turn round, but went on with his work, mending and patching till it was too dark to work any more. Then he ate up the rest of the loaf and climbed into bed. But in his sleep he dreamed that the doll came over to him and this is what she said:

> 'I can button, mend a hose,
> I can wipe a runny nose,
> If *you* won't, *I* will!'

'You're a proper fusspot, aren't you?' said the shoemaker when he woke in the morning and remembered his dream. 'I'd better get those shoes finished for Mrs Butt. Then when I take them along to her I can take the ring for the little girl as well.'

So he sat down at his bench and finished mending Mrs Butt's shoes.

Meanwhile the little boy had started playing outside the window again; the shoemaker called him in.

'Come here, boy, let me button your coat for you,' he said. 'Would you like me to wipe that nose of yours too? You can tell me if it hurts.'

' 'Course it doesn't hurt,' said the little boy, blowing into the hankie like a trumpet and waving his fingers in the air; they were quite stiff and blue with cold. 'Not so cross today, are you?' he said, and then he added: 'I'd like to help *you* now. Got any messages for me to take?'

'Well yes, I have, as a matter of fact,' said the shoemaker. 'You can take these shoes to Mrs Butt at the corner, and while you're at it you can take this ring and give it to the little girl at Crag House; she lost it here a long time ago.'

'All right!' said the little boy and ran off with the things.

It wasn't long before he was back with a message from Mrs Butt. 'Mrs Butt told me to tell you that if you want your Sunday suit cleaned and pressed she'd be glad to do it for you. And that girl at Crag House told me to thank you and ask if you'd like to go to her birthday party tomorrow—she's asked me too, wasn't that nice of her?'

The shoemaker was staring in front of him, thinking hard. 'Very nice of her, I'm sure. But, as for me, I've been on my own for so long now I wouldn't know how to behave with other folks at a party. No, I'd better not go.'

But when the little boy had gone away and the shoemaker had got into bed he dreamed again that the doll came over to speak to him:

'I always smile when folks are kind,
Not turn my back and act so blind,
If *you* won't, *I* will!'

Next morning the shoemaker remembered and said to himself, 'I shall have to go, then.' And he has never regretted it. For one thing, he wore his Sunday suit, all neatly cleaned and pressed by Mrs Butt, and, for another, he had three different kinds of cake to eat, and, most important of all, he found out that it is good for people to get together now and then and not always to be moping on their own.

Ever since that day the shoemaker's cottage has been as clean as a new pin, and the shoemaker himself whistles and sings as he goes about his housework. He makes his bed and scrubs the floor, he buttons boys' coats and wipes their noses and he mends the shoes in double-quick time.

Every now and again, when there's something he *doesn't* want to do, he takes a quick look at his doll on the chest of drawers, and he always ends up by doing whatever it is he has to do. For if *he* won't, *she* will.

Mrs Pepperpot is taken for a Witch

MRS PEPPERPOT lives in a valley in Norway, and in summertime in that part of the world the nights hardly get dark at all. On Midsummer's Eve, in fact, the sun never quite goes down, so everybody, young and old, stays up all night to dance and sing and let off fireworks round a big bonfire. And because there's magic abroad on Midsummer's Eve they sometimes see witches riding on broomsticks through the sky—or they think they do, anyway.

Now the only two people in that valley who never used to go to the bonfire party were Mr and Mrs Pepperpot. Not that Mrs Pepperpot didn't want to go, but Midsummer's Eve happened to be Mr Pepperpot's birthday as well, and on that day it was he who decided what they did. He never liked mixing with a crowd on account of that shrinking habit of Mrs Pepperpot; he was always afraid that she would suddenly turn the size of a pepperpot and disappear, leaving him standing there looking a proper fool.

But this year Mrs Pepperpot *did* go to the party, and this is how it happened.

It started the night before Midsummer's Eve. Mrs Pepperpot had been to the store and was walking slowly home with her basket on her arm. She was wondering how she could persuade her husband to go to the bonfire when suddenly she had an idea.

'I could ask him if there was something he really wanted for his birthday, and then I could say I would give it to him if he promised to take me to the bonfire party.'

As soon as she got inside the door she jumped on her husband's knee and gave him a smacking kiss on the tip of his nose.

'Dear, good hubby,' she said, 'have you got a very special wish for your birthday tomorrow?'

Her husband was quite surprised. 'Have you had sunstroke or something? How could you buy anything? Why, money runs through your fingers like water.'

'Sometimes it does, and sometimes it doesn't,' said Mrs Pepperpot, looking sly; 'there are such things as hens, and hens lay eggs and eggs can be sold. Just now I have quite a tidy sum put by, so just you tell me what you would like and the present will be laid out here on the table as sure as my name's Pepperpot.'

'Well,' he said, 'if you think you have enough money to buy that handsome pipe with the silver band that's lying in the store window, I'll promise you something in return.'

'Done!' cried Mrs Pepperpot at once, 'and the thing I want you to promise me is to take me to the bonfire party on Windy Ridge tomorrow night!'

So Mr Pepperpot had to agree and the next day Mrs Pepperpot filled her pockets with all the sixpennies, pennies and threepenny bits she had earned from the eggs and set off to the store.

'I want to buy the pipe with the silver band,' said Mrs Pepperpot, when it came to her turn to be served.

But the grocer shook his head. 'Sorry, Mrs Pepperpot,'

he said, 'but I'm afraid I sold that pipe to Peter Poulsen yesterday.'

'Oh dear,' said Mrs Pepperpot, 'I'll have to go and see if he'll let me buy it off him,' and she hurried out of the door, letting the door-bell jingle loudly as she went.

She took the shortest way to Peter Poulsen's house, but when she got there only his wife was at home.

'I was wondering if your husband would sell me that pipe he bought in the store yesterday?' said Mrs Pepperpot. 'I'd pay him well for it,' she added, and patted her pocketful of coins.

'That pipe is no longer in the house,' said Mrs Poulsen, who had a sour look on her face. 'I wasn't going to have tobacco smoke in my curtains, no *thank* you! I gave it to some boys who were having a sale; they said they were collecting money for fireworks for tonight's bonfire, or some such nonsense.'

Mrs Pepperpot's heart sank; did Mrs Poulsen know where the sale was being held?

'Up on Windy Ridge, near the bonfire, the boys said,' answered Mrs Poulsen, and Mrs Pepperpot lost no time in making her way up to Windy Ridge.

But it was a tidy walk uphill and when she got to the top she found the boys had sold everything. They were busy tidying up the bits of paper and string and cardboard boxes and carrying them over to the bonfire.

Mrs Pepperpot was so out of breath her tongue was hanging out, but she managed to stammer, 'Who got the pipe?'

'What pipe?' asked one of the boys.

'The one with the silver band that Mrs Poulsen gave you.'

'Oh that,' said the boy; 'my brother bought it. But then he tried to smoke it and it made him sick. So he got fed up with it and tied it to a long pole and stuck it at the top of the bonfire. There it is—look!'

Mrs Pepperpot looked, and there it was, right enough, tied to a pole at the very top of the huge bonfire!

'Couldn't you take it down again?' she asked the boy.

'Are you crazy?' said the boy. 'Expect us to upset the bonfire when we've got everything piled up just nicely? Not likely! Besides, we're going to have some fun with that pipe; you wait and see! But I can't stand talking now, We must go and round up the others.' And he ran off with the boys.

'Oh dear, oh dear, oh dear!' wailed Mrs Pepperpot

to herself. 'I see there's nothing for it but to climb that bonfire and get it down myself.' But she looked with dismay at the mountain of old mattresses, broken chairs, table-legs, barrows, drawers, old clothes and hats, car-tyres and empty cartons.

'First I shall have to find a stick to poke the pipe off the pole when I do get up aloft,' she thought.

Just at that moment she turned small, but for once Mrs Pepperpot was really pleased. 'Hooray!' she shouted in her shrill little voice. 'It won't take long for a little thing like me to get that pipe down now, and I don't even need to upset the bonfire!'

Quick as a mouse, she darted into the big pile and started climbing up from the inside. But it was not as easy as she had thought; climbing over a mattress she got her heel stuck in a spring and it took her quite a while to free herself. Then she had difficulty in climbing a slippery chair-leg; she kept sliding back. But at last she managed it, only to find herself entangled in the lining of a coat. She groped about in this for some time before she found her way out of the sleeve.

By now people had started gathering round the bonfire.

'All right, let them have a good look,' she thought. 'Luckily I'm too small for them to see me up here. And nothing's going to stop me from getting to the top now!'

Just then she lost her grip and fell into a deep drawer. There she lay, puffing and blowing, till she managed to catch hold of a bonnet string which was hanging over the edge of the drawer.

'Not much further to go, thank goodness!' she told herself, but when she looked down she almost fainted; it was fearfully far to the ground, and now there were crowds of people standing round, waiting for the bonfire to be lit.

'No time to lose!' thought Mrs Pepperpot, and heaved herself on to the last obstacle. This was easy, because it was an old concertina, so she could walk up it like a staircase.

Now she was at the foot of the pole and at the top was the pipe, securely tied!

'However am I going to get up there?' she wondered, but then she noticed the rim of an empty tar-barrel right next to her. So she smeared a little tar on her hands to give them a better grip, and then she started to climb the pole. But the pole and the whole bonfire seemed to be heeling a little over to one side, and when she looked down she nearly fell off with fright: *the boys had lit the bonfire!*

Little flames were licking up round the mattresses and the broken furniture.

Then people started cheering and the children chanted: 'Wait till it gets to the pipe at the top! Wait till it gets to the pipe at the top!'

'Catch me waiting!' muttered Mrs Pepperpot. 'I've got to get there *first*!' and she climbed on up till her hands gripped the stem of the pipe. Down below she could hear the children shouting:

'Watch the flames when they reach the pole! There's a rocket tied to the pipe!'

'Oh, good gracious!' cried Mrs Pepperpot, clinging on for dear life. BANG! Up into the cold night sky shot the rocket, the pole, the pipe *and* Mrs Pepperpot!

Round the bonfire everyone suddenly stopped shouting. A thin woman in a shawl whispered to her neighbour:

'I thought I saw someone sitting on that stick!'

Her neighbour, who was even thinner and wore two shawls, whispered back, 'It could have been a witch!' and they both shuddered. But from behind them came a man's voice:

'Oh, it couldn't be her, could it?' It was Mr Pepperpot who had just left off working and had taken a ride up to the mountain to have a look at the bonfire. Now he swung himself on his bicycle again and raced home as fast as he could go, muttering all the way, 'Let her be at home; oh, please let her be at home!' When he reached the house and opened the door his hand was shaking.

There stood Mrs Pepperpot, quite her normal size and with no sign of a broomstick. She was decorating his birthday cake and on the table, neatly laid on a little cloth, was the precious pipe with its silver band.

'Many Happy Returns of the Day!' said Mrs Pepperpot. 'Come and have your meal now. Then you can put

on a clean shirt and take your wife to dance all night at the bonfire party!'

'Anything you say!' said Mr Pepperpot; he was so relieved she hadn't gone off with the witches of Midsummer's Eve.

The Little Mouse who was Very Clever

THERE WAS once a little mouse called Squeak, who sat behind the door in his mousehole, waiting for his mother to come home from the larder with some food.

Suddenly there was a knock at the door.

'Who is that?' asked little Squeak.

'Peep, peep, let me in, it's your mother,' said a voice outside, but it didn't sound a mousy sort of voice.

'Why are you knocking at our door?'

'I forgot to take the key,' said the voice.

'We don't use a key for this door,' said the little mouse.

'It's bolted on the inside and I'm not opening it for *you*!'

Whoever it was went away and Squeak waited for some time before there came another knock.

'Who is that?' asked Squeak.

'It's Peter,' said a voice, but it didn't sound like Peter, the little boy who lived in the house.

'Why are you knocking at our door?'

'Because you've been using my shawl for your bed,' said the voice.

'If you're a little boy, you won't want a shawl,' said Squeak. 'I'm not opening the door for you!'

After he had waited some time there was another knock.

'Who is that?' he asked.

'Peter's mother,' came the reply; 'let me in.'

'Why are you knocking at our door?'

'Because I think you must have hidden my braces down your hole,' said the voice, which sounded rather purry for a lady.

'If you are Peter's mother,' said little Squeak, 'I'm sure you don't need braces. I'm not opening the door for *you*!'

Again he had to wait a long time before the next knock.

'Who is that?' said Squeak.

'This is Peter's father,' said a voice that was very croaky. 'Let me in at once!'

'Why are you knocking at our door?' asked the little mouse.

'My tail went in the honey-pot and I want you to clean it.'

'If you are Peter's father,' said Squeak, 'why have you got a tail? I'm not opening the door for *you*!'

After that there was a long, long silence, and then at last there was a tiny little knock.

'Who is that?' asked Squeak.

'It's your mother, darling.'

But the little mouse wanted to make quite sure:

'Why are you knocking at our door?'

'Because I want my clever Squeak to unbolt it as he always does,' said the voice. Then he knew it was his mother, and the little mouse, who was too clever to be tricked by a cat with many voices, unbolted the door and let his mother in.

And because he'd been so clever his mother gave him an extra large lump of cheese for his supper.

Mrs Pepperpot's Birthday

It was Mrs Pepperpot's birthday, so she had asked her neighbours in to coffee at three o'clock. All day she had been scrubbing and polishing, and now it was ten minutes to three and she was putting the final touches to the strawberry layer cake on the kitchen table. As she stood balancing the last strawberry on a spoon, she suddenly felt herself shrinking, not slowly as she sometimes did, but so fast that she didn't even have time to put the strawberry on a plate. It rolled on to the floor and Mrs Pepperpot tumbled after it. But she quickly picked herself up and jumped into the cat's basket. Puss was a bit surprised, but allowed her to snuggle down with the kittens. In her black-and-white-striped skirt and white blouse, she hoped the guests would take her for one of the cat-family, until the magic wore off and she could be her real size again. For you may remember that Mrs Pepperpot never liked anyone to see her when she was tiny.

There was a knock at the door, and, when it wasn't

answered, Sarah from South Farm walked into the little front hall, carrying a huge bunch of lilac.

'Many Happy Returns of the Day!' said Sarah. There was no reply, so she peered into the kitchen, thinking Mrs Pepperpot might be in there, though, naturally, she didn't look in the cat-basket. Somehow she managed to knock over the flower-vase on the hall table, and the water spilled on the tablecloth and on to the floor.

'Oh dear, oh dear!' thought Sarah. 'I shall have to mop that up before anybody notices.'

But at that moment there was another knock at the

door. So Sarah ran into the kitchen and hid in a cupboard.

In came Norah from North Farm, and she was carrying a very nice tablecloth.

'Many Happy Returns!' she said, but, as she got no answer, she looked round for Mrs Pepperpot, and her parcel swept the vase off the table on to the floor.

'That's bad!' thought Norah. 'I must put it back before anybody comes.'

But before she could do it there was another knock on the door and Norah hurried into the bedroom and crept under the bed.

Esther from East Farm came in, carrying a handsome glass bowl for Mrs Pepperpot. When she had said 'Many Happy Returns!' and no one answered, she walked

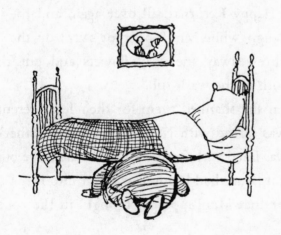

straight into the living-room. Carrying the bowl in front of her, she didn't notice the vase on the floor and put her foot straight on it. There was a nasty crunch and there it lay, in smithereens!

'Goodness gracious, what have I done?' thought Esther. 'Perhaps if I hide behind this curtain no one will know who did it!' So she quickly wrapped herself in one of the curtains.

At that moment the clock struck three, and the magic wore off; there was Mrs Pepperpot as large as life, walking through the kitchen. 'Coo-ee!' she called. 'You can all come out now!'

So Sarah stepped out of the cupboard, Norah crawled from under the bed, while Esther unwrapped herself from the curtain in the living-room.

At first they looked a bit sheepish, but then they said 'Many Happy Returns!' all over again and they had a good laugh, while Mrs Pepperpot swept up the broken vase, threw away the dead flowers and put the wet tablecloth in the wash-tub.

Then she thanked them for their fine presents; the table was spread with Norah's tablecloth, Esther's glass bowl was filled with fresh water, and the huge bunch of lilac that Sarah had brought was put into it.

After that Mrs Pepperpot brought in the coffee and

cakes and they all sat down to enjoy themselves. But on the strawberry layer cake there was one strawberry missing.

'You see,' said Mrs Pepperpot, when they asked her what had happened to it, 'my *first* visitor this afternoon was the little old woman who shrinks, and she was so tiny today that one strawberry was all she could manage to eat. So I gave her that and a thimble of milk to wash it down.'

'Didn't you ask her to stay, so that we could see her?' asked Sarah, for they were all very curious about the little old woman who shrank, and nobody thereabouts had ever seen her.

'She was sorry, she said, but she was in a tearing hurry; she had some business with a mouse, her night watchman or something. But she told me to tell you she did enjoy our little game of hide-and-seek!'

The Dancing Bees

You have probably been told that bees are very busy, hardworking creatures, collecting honey from the flowers from morning till night all through the summer to store it up for food in winter.

But now you shall hear about a she-bee who spent her time in quite a different way; she did nothing but dance. And what gave her this curious idea? Well, wait till I tell you.

One evening in June this she-bee was on her way home to the hive from the meadows, but she hadn't found any honey that day. So when she saw a beautiful flower right below her she swooped down and settled on it. But the flower was swinging round because it was fixed to a hat, and the hat was on the head of a girl, and the girl was dancing with her young man to the band in People's Park.

At first the little she-bee was disappointed to find no honey in the flower, but soon she forgot all about it; it

was such a delicious feeling to sway up and down in time to the music.

'It's like hearing a thousand bumble-bees humming over a field of clover,' she murmured happily as she swayed round and round.

'Yes, isn't it?' said a voice quite close to her. And there, on the boy's hat, sat another bee!

'I shouldn't talk to strangers, I know,' said the little she-bee, 'but this music is making me so giddy I don't really know what I'm doing.'

'Nothing wrong in talking, surely?' said the other bee, who was, of course, a he-bee. 'Allow me to say that you dance superbly, and it's not so easy to hang on to a flower that is swinging round and round. I know what I'm talking about; there were two young lady bees here earlier this summer and I tried to dance with them, but they both fell off after just a few turns. I always hold on to the young man's hat-brim; in any case, this is my second season here, so I've had plenty of practice. Ah, the music is stopping; that means it's time for refreshments. Just stay where you are.'

The girl and the young man moved over to the refreshment tent and the bees went with them on their hats.

'Do you live far from here?' asked the he-bee.

'It's quite a long flight,' said the she-bee, 'but I simply daren't let anyone see me home, as we have such a very strict doorkeeper.'

'I quite understand,' he replied, 'but perhaps you will permit me to fetch you a little drink of fruit-juice?'

He flew down on to the table and a moment later brought back a drop of fruit-juice on one of his wings. 'Excuse my wing, but it is the only way I can manage it.'

'Very nice of you to bother; thank you very much,' said the little she-bee. 'Aren't you having any yourself?'

'Not for me, thanks, I'm driving—at least, it's not exactly *me* driving, but I daren't take any risks, all the same. I shall have to get back to my own hat now, I see they're starting the show.'

The little she-bee would have liked to ask what a 'show' was, but she felt shy, so she just rubbed her wings with her feet.

'Tonight there's a fellow with a guitar—forgotten his name, but he's very famous,' the he-bee told her. 'One young lady bee told me she once sat on his guitar while he played "Rocking round the Clock", and she'll never forget it as long as she lives, she said.'

The little she-bee listened while the man on the platform sang and played his guitar, and very funny he looked. First he sang quite quietly, but suddenly he went into shrieks and sobs and then he fell right down on his knees and shuffled off the platform backwards.

'What happened? Did a bee sting him?' asked the she-bee.

'No, no, that was the high spot,' answered her new friend.

'The high spot? Oh yes, I've heard that high spots can sting very badly—the poor man!' And the little she-bee felt quite sorry for the man. But then the music began once more. 'Isn't it wonderful!' she cried, and clapped her front pair of feet.

They sailed out on the dance-floor again, she on her flower and he on his hat-brim.

'I'm afraid that's the last dance tonight,' he said. 'I can see they're beginning to put out the lights.'

'How sad! I shall have to hurry home now,' said the little she-bee, 'but thank you so much for a lovely evening.'

'You're welcome! Perhaps we could meet another time?'

'That would be very nice,' she answered.

When she got back to the hive the doorkeeper would not let her in at first because she had no honey with her. But in the end he relented.

Next evening she went dancing again. But this time

she was careful to hide a little honey in a tree-stump close to the hive, so that she could pick it up on her way home after the dance.

The he-bee on his hat-brim was very pleased to see her. 'I have been thinking about you all day,' he said.

'Very kind of you, I'm sure,' she said, looking very demure.

Like the evening before, there was music, there was dancing, the he-bee brought her fruit-juice, and then there was the show. Afterwards he persuaded her to take a little ride in the young man's car, but when the he-bee asked if he might sit on *her* hat the little she-bee said firmly, 'No.'

Later, when she got back to the hive, she had no difficulty in getting past the doorkeeper, as she had remembered her store of honey in the tree-stump.

And so it went on every evening all through the summer, and no one in the hive knew anything about it. But one day, while the she-bee was flying round looking for honey, she forgot herself; instead of her ordinary buzzing song she started to sing 'Waltzing Matilda'.

'Whatever are you buzzing?' asked the bee who was flying nearest to her.

The little she-bee got quite confused. 'It's—it's only a little waltz tune I heard,' she stammered. Then, of course, she had to explain *where* she had heard it. But this was just as well, for she could not have kept the secret to herself much longer.

As the day went on other bees kept flying up to her wanting her to tell them about the dancing, the music and the show, and about her dancing partner on the hat-brim.

'Can't you take us along with you tonight?' they all pleaded.

'Not tonight,' she said, 'but perhaps tomorrow, if my partner gives his permission.'

That evening was as delightful as all the rest, but the young man didn't have his car with him, so he and the girl and the bees all went for a walk in the wood instead.

'It is really just as pleasant, don't you think, to go for a walk on such a lovely evening?' said the he-bee. 'And if I'm not mistaken we're heading for that little bank with the bluebells on it.' He was right; the young man and the girl sat down on the flowery bank, and so did the two bees.

The young man spoke softly to the girl, and she bent her head so far forward that the bee had to hold on tight not to fall off the flower.

'Why does she bend her head like that?' she asked.

'He is proposing to her,' said the he-bee. 'You had

better let me come over and hold you so you don't fall off.' And he took off from the hat-brim and made an elegant landing on the flower. Then he put his wing protectingly round the little she-bee. This time she did not protest.

'What does "propose" mean?' she asked, though deep inside she had a feeling she knew what it meant already.

'He is asking if she will be his wife and stay with him for ever.'

'Oh, how wonderful!' cried the little she-bee before she could stop herself.

'Do you really think so? Then I shall propose to you this very minute,' said the he-bee, 'even though this isn't the usual way for bees to behave, I'm sure. But it's difficult, you will agree, not to be carried away in these beautiful surroundings and hearing so much love-talk!'

'Oh yes!' sighed the little she-bee, and hung her head bashfully just like the young girl.

The next day there was a terrific hullabaloo in the hive. The bees who knew about the dancing had told a lot more bees, and now it had got all round the hive. There they were, a great black swarm of them, thronging round the entrance, all demanding to go to People's Park like the little she-bee.

'Show us the way!' they kept crying. 'Show us the way and we'll follow you!'

The little she-bee hesitated at first, but then she said: 'All right, I will. You can all come with me.' Secretly she was longing for the music and for her partner in the park.

So the whole swarm rose in the air and flew off after the she-bee. But this was in the middle of the day and there was not a soul to be seen in People's Park when they got there. No sign of the flower on the hat, or, for that matter, of the girl and the young man and the he-bee. The she-bee was lost and she knew it, but she couldn't stop now.

On, on she flew, and the swarm after her, over fields and meadows, over roads and hedges, until quite suddenly the little she-bee caught sight of something hanging on the branch of an apple tree.

It was the hat with the flower on it! She was saved!

'Come on, all of you!' she buzzed as loudly as she could. She flew to the flower and clung there. And all the other bees clung to her, till they looked like a huge black cluster of grapes shining in the sunshine.

Someone came into the orchard; it was the girl. When she saw what was happening to her hat she gave a shriek of alarm. She shouted out to someone inside the house, and out came the young man with a great big net. He put it carefully over the swarm of bees and carried it over to an empty hive. There he shook the net so that they all tumbled in, and then he put the lid on firmly.

As luck would have it, the he-bee was still perched on the young man's hat-brim, so he was able to slip into the hive with the rest of them.

After they had time to settle down in their new home the bees made the little she-bee their queen. But, from what I've heard, she was not a bit kind to the he-bee when he became her husband. And if he or any of the other bees ever dared to dream about going dancing in People's Park they soon discovered that it was the queen who called the tune from now on; their dancing days were over for good.

How the King Learned to Eat Porridge

THERE WAS once a king who was wise and good and mighty and rich, but he had one fault. Not a very big fault, mind you, but annoying enough. And do you know what it was?

He would not eat porridge.

Now this wouldn't have mattered so much if he had kept it to himself. After all, a king has so many other things to eat. But he was an honest king, so he could not lie, and he told both the government and the people that he did not like porridge.

'I detest porridge!' he declared. 'It tastes of glue and tufts of wool. Ugh!'

And when the people of his realm heard that they wouldn't eat porridge either. They wanted to be like the king—especially the children, who thought it was wonderful not to eat porridge, and whooped with delight. On the national flag-day they walked in the procession carrying big banners saying 'Long Live the King! Down with Porridge!'

But far up in the mountains there lived a peasant with his daughter, and they had only one tiny field of their own. On this field they grew oats, and their whole livelihood depended on it. So when everyone stopped buying oats they made less and less money.

'This can't go on!' said the peasant's daughter one day. 'People not eating porridge like this. I shall have to go to the palace and give that stupid king a piece of my mind!'

'Take care, my child, you might make the king angry,' said her father.

'He'd better take care himself,' she answered, 'or *I* might get angry.' And she slung a bag of oatmeal over her shoulder and set off for the palace.

In the palace garden she met a man, but the country girl did not know he was really the king.

'Where are you off to?' asked the man.

'I'm going to the palace to teach the king to eat porridge,' she answered.

This made the man laugh. 'If you can do that,' he said, 'you're a very clever girl!' Then he went into the royal kitchen and told the royal cook that he could take a day off, as a country girl had come to town to teach him, the king, to eat porridge.

Next day, when the king was at breakfast, the

peasant's daughter brought in a big steaming bowl of porridge. She was a little taken aback when she saw who the man was she had spoken to in the garden. But the king smiled at her and said:

'Come along now, you must teach me to eat this porridge of yours.'

'Oh no,' she said, 'not after making such a fool of

myself, I couldn't. Anyway, I don't suppose you even know which hand to hold the spoon in?'

'I think I do,' said the king, and picked up the spoon in his right hand.

'Just as I thought!' said the girl. 'Never in all my life have I seen anyone eat porridge with the right hand!'

'All right,' said the king, 'I'll try with my left hand.'

'That's better,' said the girl, 'but I've never seen anyone with a bowl of porridge on the table while they sat on a chair to eat it. Get up on the table and put the bowl on the chair in the proper manner!'

The king was beginning to enjoy himself. First he got

up on the table and then he tried to set the bowl on the chair. But this was not so easy; he nearly lost his balance several times before he managed it.

'There, now you can start eating,' said the girl.

The king held the spoon in his left hand and bent down to the bowl. It was awkward, but he succeeded at last in getting a spoonful up to his mouth and swallowing it.

'Just a minute,' said the girl, 'you have to hold your left ear with your right hand while you're eating.'

'I expect I can manage that too,' said the king, and balanced another spoonful up to his mouth.

'That's only two spoonfuls,' she said. 'Wait till you get to the bottom of the bowl; that's when it gets really difficult.'

'Don't worry, I'll do it!' said the king, and, do you know, he didn't give in till he had scraped the bowl clean!

'There you are!' he shouted proudly.

'There you **are yourself**!' said the girl. 'Now I have taught the king to eat porridge!'

At first the king was a little put out that she had tricked him in this way. But then he had to laugh.

'You're a very cunning little girl,' he said. 'I almost think I'll marry you—that is, if you're willing?'

'I might as well,' said the girl.

*   *   *   *   *

So they were married, and at the wedding a huge pot of hot, steaming porridge was set up in the middle of the market place, and the king commanded his people, including the whole court and all his ministers, to come and learn how to eat porridge.

The queen stood on a high tribune to teach them, and they all laughed and enjoyed themselves like anything. But when they had had their lesson the queen said:

'Now I know you can all eat porridge; so from this day on you may sit on your chairs and put your porridge bowls on the table and eat it the way you do any other food.'

But many people, especially the children, preferred the way the queen had taught them and went on doing it—secretly.

Mrs Pepperpot Turns Fortune-Teller

EVERY MORNING when Mr Pepperpot goes off to work
Mrs Pepperpot stands at the window and watches him
till he disappears round the bend to the main road. Then

she settles down in the chair by the kitchen table, picks up her empty coffee cup and starts reading her fortune in it.

Now you probably didn't know that Mrs Pepperpot could read fortunes in a cup. Well, she can; she can tell from the way the coffee grounds lie what road she will take that day and whether she will meet joy or sorrow before nightfall. Sometimes she sees the shape of a heart in the cup and that means she will have a new sweetheart. But that makes Mrs Pepperpot laugh, for to her it means she will probably get a new pet to look after—perhaps a poor little bird with a broken wing or a stray kitten on her doorstep, getting tamer and tamer as it laps up the food and milk she gives it.

But if the grounds form a cross she knows she must watch her step, for that means she will break something; it could be when she is washing up or when she is scrubbing the floor. If she sees a clear drop of coffee running down the side of the cup that means she will hurt herself in some way and will need not only a bandage but maybe a doctor as well. And so it goes on; there are many more signs that she can read, but she only does it for herself, never for other people, even if they ask her. It's just an amusement, she says, something to while away the time when she is at home alone all day.

Well, this day—it was a Friday too—Mrs Pepperpot had planned to give the house a good clean out and then she was going to bake a cake for Mr Pepperpot. Apart from that she was just going to take it easy for a change. So, when she had watched her husband turn the corner, she picked up their two coffee cups and was just about to put them in the sink. But then she stopped herself.

'There, what am I doing? I nearly forgot to have a look at my fortune for today!' So she took one of the cups back to the table and sat down. 'Let's see, now,' she said and turned the cup round and round in her hand. 'Oh dear, oh dear!' she exclaimed, 'what's this I see? A big cross? I shall have to mind how I go today and no mistake!'

At that very moment she shrank, and in no time at all she was no bigger than the coffee cup and both she and the cup fell off the chair on to the floor.

'That was a bit of a come-down!' she said, and felt both her arms and her legs to see if there were any bones broken. But when she found she was still all in one piece she lay still for a moment, not daring to look at the cup. For it was one of her best ones, and she was sure it must have been broken by the fall.

At last she said to herself, 'I suppose I shall have to have a look.' And when she did she found to her

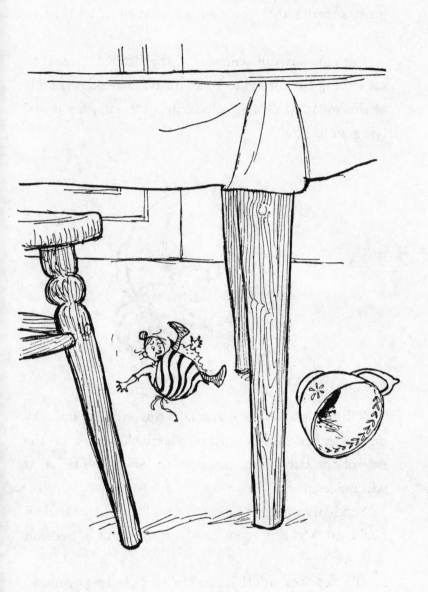

great surprise that the cup was not even cracked or chipped.

But she was still worried. 'If that isn't it there'll be something else for me to break today,' she said miserably as she squatted down to look into the cup, for it was lying on its side.

'Oh me, oh my!' she cried. 'If this isn't my unlucky day!' She had caught sight of a large clear drop on the side of the cup. 'This means tears, but I wonder what will make me cry?'

Suddenly there was a loud BANG! inside the kitchen cupboard. Mrs Pepperpot nearly jumped out of her skin with fright.

'There! Now isn't that just like Mr P., setting a mouse-

trap in the cupboard, although he knows it's not necessary
now that I have a mouse for a night watchman. It's only
now and again that a baby mouse gets into the cupboard
by mistake—before he's learned the mouse rules. It's not
as if they *mean* to do any damage, so it's silly to take any
notice. I wonder if I dare open the door a little to see
what's happened? I suppose I'd better; the little thing
might just have caught its tail and I could free it. But, of
course, it might be worse than that; the cup said tears,
and tears it will be, no doubt!'

So Mrs Pepperpot went over to the cupboard and
pulled gently at the door. But she closed her eyes to keep
back the tears which were ready to come at any moment.
When she had the door opened enough to look in she
opened first one eye, then the other, and then she flopped
down on the floor, clapping her hands on her knees, and
burst out laughing.

Right enough, the trap had snapped, but there was
nothing in it. Instead two baby mice were happily playing
just beside it with two empty cotton-reels. Mrs Pepperpot
thought it was the funniest sight she'd ever seen.

'Hullo, Mrs Pepperpot!' squealed one of the baby
mice. 'Have you shrunk again?'

'We hoped you would!' said the other one, ' 'cos my
brother and I had never seen you small before, and

Granny said we could come in here and have a peep—just in case you shrunk. We weren't being naughty—just playing cars—and then we bumped into that nasty thing which went snap over there.'

'Will you play with us?' asked the first little mouse. 'You sit in the car and we'll pull you along.'

And when Mrs Pepperpot looked closer she saw that the baby mice had fixed a matchbox over the cotton-reels and the whole contraption really moved.

'Let's go!' shouted Mrs Pepperpot, and jumped into the box.

So they played at cars, taking turns to sit in the matchbox, and Mrs Pepperpot laughed while the baby mice squealed with delight, till, all of a sudden, they heard a scratching sound above them.

'That's enough, children!' called granny mouse,

whose head had appeared in a hole in the back wall. 'The cat's on top of the cupboard and the door is open!'

Before you could say 'knife' the two baby mice had disappeared through the hole, squeaking 'Thanks for the game!' as they went.

'Thank you!' said Mrs Pepperpot, and stepped out of the cupboard to see what that cat was up to.

There she was, standing on top of the cupboard waving her tail expectantly when Mrs Pepperpot came out. But Mrs Pepperpot was not standing any nonsense; she shouted at the cat: 'What are you doing up there? You get down at once or I'll teach you a lesson as soon as I grow again! Maybe it's you who are going to break something for me today? Yes, I can feel it in my bones. I know if I have another look in that cup there'll be more calamity there for me.'

By now she was just as worked up as she had been before her little game with the mice. But she couldn't re-sist having another look in the cup. 'Goodness gracious!' she cried. 'It's just as I thought, doctor and bandages, ambulances and everything! As if I hadn't trouble enough already! Down you get, cat, and make it quick!'

'All right! Keep your hair on,' said the cat. 'I was only doing my duty when I heard a suspicious noise in the cupboard. I'm coming down now.'

'Mind how you go! Be careful! I don't want anything broken. I'll stand here below and direct you,' said Mrs Pepperpot.

'Anybody would think I'd never jumped off a cupboard before, and I'm not in the habit of breaking things,' answered the cat, as he made his way gingerly past a big china bowl. But just on the edge of the cupboard lay a large pair of scissors, and neither the cat nor Mrs Pepperpot had seen them.

Mrs Pepperpot was busy with her warnings: 'Mind that bowl!' she shouted, standing right beneath those scissors.

The cat was being as careful as she could, but her tail brushed against the scissors, sending them flying, point downwards, to the floor. There they stood quivering!

Mrs Pepperpot had just managed to jump out of the way, but now she was too frightened to move. 'So that was it!' she stammered at last. She felt herself all over again, for this time she was *sure* she must be hurt. But she couldn't find as much as a scratch!

A moment later she was her normal size. So she pulled the scissors out of the floor and lifted the cat down out of harm's way. Then she set to work cleaning the house and just had time to bake her cake before she heard her husband at the door.

But what a state he was in! The tears were pouring out of his eyes because of the bitter wind outside, and anyway he had a bad cold. One hand he was holding behind his back: he had fallen off his bicycle, had broken his cycle lamp and cut his hand on the glass!

As she hurriedly searched for something to tie round his hand, Mrs Pepperpot thought how odd it was that it was Mr Pepperpot who had tears in his eyes; it was Mr Pepperpot that had broken something, and had hurt himself so that he had to have a bandage on. Very odd indeed!

But if you think this cured Mrs Pepperpot of reading her fortune in a coffee cup you are very much mistaken. The only thing is, she *does* take more care not to pick up the wrong cup and read her husband's fortune instead of her own.

The Fairy-Tale Boy

ONCE UPON A TIME there was a little boy who said to his mother, 'I want to go out into the world and kill a dragon, and then I'll bring you back all the gold.'

His mother answered: 'But what about the princess? Won't you set her free as well, pet?'

'I suppose so,' said George, for that was his name, 'as long as I don't have to marry her; 'cos you know I don't like girls.'

'Very well, dear; good luck!' said his mother.

'Oh, but you mustn't just say "good luck!" You have to give me food and drink for the journey. I'm going to sit down by the side of the road and start eating, and then an old woman will come or a dwarf, or something like that, and they'll be hungry, and I'll give them all my food, 'cos then they'll help me, you know, to find the way to the dragon and all that.'

'All right, pet, you shall have some food for the journey,' said his mother, and she put three sandwiches in a tin and put them in a little knapsack with a bottle of milk.

So George went out into the world to kill the dragon. And when he had walked a little he came to a road crossing and there he sat down and opened his knapsack and took out a sandwich.

'I 'spect an old woman will come along in a minute, and she'll ask me for food. Then afterwards she's sure to show me the way to the dragon,' he said to himself.

No sooner had he thought this than an old woman

did come round the corner, walking slowly towards him. When she reached the place where he was sitting she stopped and fixed her eyes on George's sandwich.

'Good afternoon,' said George, and held out his hand with the sandwich in it.

'Good afternoon, my boy,' said the woman; 'you shouldn't play with your food like that.'

'I'm *not* playing with my food,' said George indignantly. 'I'm giving it to you. Aren't you hungry?'

'Of course not,' said the woman. 'I have my lunch before I go out, and so should you, young fellow, and not drag it about with you, leaving dirty bits of paper behind and egg-shells and I don't know what. Ugh!' And with that she walked off and left George holding his sandwich.

George sat staring after her for a bit.

'Oh well,' he said, 'I suppose there'll be others coming along. She wasn't the right kind of fairy-tale woman, anyway. I'd better eat this sandwich myself while I'm waiting.'

While he was munching the bread he heard a scuffling noise behind the hedge. It was a squirrel.

'Hullo, squirrel! Would you like something to eat?' said George, and threw the second sandwich into the hedge. But the squirrel took fright and bounded away towards a wood.

'Perhaps it wants me to follow it,' thought George. So he picked up his knapsack and walked after the squirrel to the wood. The squirrel, meanwhile, stayed quite still, as if it were really waiting for him, and George was now quite certain that it was going to show him the way to the dragon. When he got quite close the squirrel started leaping from branch to branch and George began to run, stumbling over sticks and stumps, further and further into the wood, until at last they came to a little stream. Here the squirrel jumped lightly from a tree on one side of the stream to another tree on the other side. But George stopped; the stream was too deep for him to wade across, and he was afraid he might drown. So he very sensibly stayed where he was.

Instead he sat down by the edge of the stream and took out his last sandwich. Just then he saw a fish swimming by.

'P'raps you're the one who is going to help me,' he said, and threw the last sandwich into the water. But the fish dodged under a stone and the sandwich floated away down the stream.

George jumped up and followed the bobbing sandwich to see where it would go. Soon the stream widened and ran into a pool, and there on the bank stood a big boy watching a fishing-line.

'Don't make so much noise!' said the big boy. 'You'll frighten the fish away!'

'Sorry!' said George. 'Would you like some milk?' he asked.

'Milk? Are you off your head? How d'you expect me to drink milk while I'm fishing? You scram out of here and be quick about it!'

Well, there was nothing for it; George had to make himself scarce, but he was beginning to feel a little sad. His adventure wasn't turning out quite as he had planned. He sat down in a patch of bilberries and drank up his bottle of milk himself. Then he lay down and took a little nap.

When he woke up there was a little girl standing there, watching him.

'Hullo!' said George.

'Hullo!' said the little girl. 'You looked so nice, lying there asleep; just like my cousin! What a smart sandwich tin you've got; I haven't got one like that with Donald Duck on it.'

'Got it for my birthday,' said George. He was suddenly feeling very hungry, but there was nothing left to eat.

'Are you hungry?' asked the little girl. 'I can give you some food, if you like. I'm bilberrying, you see, so I've brought a picnic lunch—look! Sausages and strawberry jelly; it's jolly good!'

'Thank you very much,' said George, and he ate up two sausages and all the jelly that was left.

The little girl stood watching him, and when he had finished she said, 'Now you will have to save me from the dragon.'

'Dragon?' George gulped and the last bit of jelly went down the wrong way.

'Of course,' she said. 'When you've eaten the princess's food you must kill the dragon. It is lying just down there in the bushes.'

'All right, I'm ready!' said George in his most manly

voice. He was a *little* afraid when he started walking over to the bushes. But he picked up a good stout stick—sorry, I mean sword—and smacked the dragon right on the nose. Wham! Afterwards he hacked it into several pieces and then he jumped on them to make quite sure it was dead.

'It's quite safe now, Princess!' he said to the little girl. 'You can come out now!'

'Oh, thank you, St George!' she said. 'Now you can

marry me if you like—unless I marry my cousin, of course.'

'You marry your cousin,' said George. 'I don't mind.'

'All right. But you can come home with me if you like, and have some more jelly.'

'No thanks, Princess,' said George. 'You see, I have to look for the gold now that I've killed the dragon.'

And do you know? Under the tree-stump—sorry, I mean the dragon—George found a heap of beautiful pebbles that shone in the sun like silver when he ran them through his hands. George put them all in his knapsack and was very pleased with his treasure even though it wasn't gold.

After all, you can't expect to find gold under the first dragon that you kill.

The Ski-Race

MRS PEPPERPOT has done a lot of things in her life, and most of them I've told you about already. But now I must tell you how she went ski-racing one day last winter.

*   *   *   *   *

Mr Pepperpot had decided to go in for the annual local ski-race. He had been a pretty good skier when he was young, so he said to Mrs Pepperpot:

'I don't see why I shouldn't have a go this year; I feel more fit than I have for many years.'

'That's right, husband, you do that,' said Mrs Pepperpot, 'and if you win the cup you'll get your favourite ginger cake when you come home.'

So Mr Pepperpot put his name down, and when the day came he put on his white anorak and blue cap with a bobble on the top and strings under his chin. He slung

his skis over his shoulders and said he would wax them
when he got to the starting point.

'Righto! Best of luck!' said Mrs Pepperpot. She was
already greasing the cake-tin and stoking the stove for
her baking.

'Thanks, wife,' said Mr Pepperpot and went off. It was
not before he had turned the corner by the main road
that Mrs Pepperpot caught sight of his tin of wax which
he had left on the sideboard.

'What a dunderhead that man is!' exclaimed Mrs
Pepperpot. 'Now I shall have to go after him, I suppose;
otherwise his precious skis are more likely to go backwards than forwards and there'll be no cup in this house
today.'

So Mrs Pepperpot flung her shawl round her
shoulders and trotted up the road as fast as she could
with the tin of wax. When she got near the starting
point there was a great crowd gathered. She dodged in
and out to try and find her husband, but everyone

seemed to be wearing white anoraks and blue caps. At last she saw a pair of sticks stuck in the snow with a blue cap hanging from the top. She could see the initials P.P. sewn in red thread inside.

'That must be his cap,' thought Mrs Pepperpot. 'Those are his initials, Peter Pepperpot. I sewed them on myself in red thread like that. I'll just drop the wax in the cap; then he'll find it when he comes to pick up his sticks.'

As she bent forward to put the wax in the cap she accidentally knocked it off the stick and at that moment

she shrank so quickly that it was she who fell into the cap, while the tin of wax rolled out into the snow!

'No harm done,' thought Mrs Pepperpot; 'when he comes along he'll see me in his cap. Then he can put me down somewhere out of the way of the race. And as soon as I grow large again I can go home.'

But a moment later a big hand reached down, snatched up the cap and crammed it over a mop of thick hair. The strings were firmly tied and Mrs Pepperpot was trapped!

'Oh well!' she thought. 'I'd better not say anything before the race starts.' For she knew Mr Pepperpot hated to think anybody might get to know about her shrinking.

'Number 46!' she heard the starter shout, 'on your mark, get set, go!' And Number 46, with Mrs Pepperpot in his cap, glided off to a smooth start.

'Somebody must have lent him some wax,' she thought; 'there's nothing wrong with his skis, anyway.' Then from under the cap she shouted, 'Don't overdo it, now, or you'll have no breath left for the spurt at the end!'

She could feel the skier slow up a little. 'I suppose you know who's under your cap?' she added. 'You had forgotten the wax, so I brought it along. Only I fell into your cap instead of the wax.'

Mrs Pepperpot now felt the skier's head turn round to see if anyone was talking to him from behind.

'It's me, you fool!' said Mrs Pepperpot. 'I've shrunk again. You'll have to put me off by the lane to our house—you pass right by, remember?'

But the skier had stopped completely now.

'Come on, man, get a move on!' shouted Mrs Pepperpot. 'They'll all be passing you!'

'Is it . . . is it true that you're the little old woman who shrinks to the size of a pepperpot?'

'Of course—you know that!' laughed Mrs Pepperpot.

'Am *I* married to *you*? Is it *my* wife who shrinks?'

'Yes, yes, but hurry now!'

'No,' said the skier, 'if that's how it is I'm not going on with the race at all.'

'Rubbish!' shouted Mrs Pepperpot. 'You *must* go on! I put a cake in the oven before I went out and if it's scorched it'll be all your fault!'

But the skier didn't budge.

'Maybe you'd like me to pop out of your cap and show myself to everybody? Any minute now I might go back to my full size and then the cap will burst and the whole crowd will see who is married to the shrinking woman. Come on, now! With any luck you may just do it, but there's no time to lose; HURRY!'

This worked; the skier shot off at full speed, helping himself to huge strides with his sticks. 'Fore!' he shouted as he sped past the other skiers. But when they came to the refreshment stall Mrs Pepperpot could smell the lovely hot soup, and she thought her husband deserved a break. 'We're well up now,' she called. 'You could take a rest.'

The skier slowed down to a stop and Mrs Pepperpot could hear there were many people standing round him. 'Well done!' they said. 'You're very well placed. But what are you looking so worried about? Surely you're not frightened of the last lap, are you?'

'No, no, nothing like that!' said the skier. 'It's this cap of mine—I'm dead scared of my cap!'

But the people patted him on the back and told him not to worry, he had a good chance of winning.

Under the cap Mrs Pepperpot was getting restless again. 'That's enough of that!' she called. 'We'll have to get on now!'

The people who stood nearest heard the voice and wondered who spoke. The woman who ladled out the soup said, 'Probably some loud-speaker.'

And Mrs Pepperpot couldn't help laughing. 'Nearer the truth than you think!' she thought. Then she called out again, 'Come on, husband, put that spurt on, and let's see if we can make it!'

And the skis shot away again, leaping many yards each time the sticks struck into the snow. Very soon Mrs Pepperpot could hear the sound of clapping and cheering.

'What do we do now?' whispered the skier in a miserable voice. 'Can you last another minute? Then I can throw the cap off under the fir trees just before we reach the finishing line.'

'Yes, that will be all right,' said Mrs Pepperpot. And, as the skis sped down the last slope, the strings were untied and the cap flew through the air, landing safely under the fir trees.

When Mrs Pepperpot had rolled over and over many times she found herself growing big once more. So she got up, shook the snow off her skirt and walked quietly home to her house. From the cheering in the distance she was sure her husband had won the cup.

The cake was only a little bit burnt on the top when she took it out of the oven, so she cut off the black part and gave it to the cat. Then she whipped some cream to put on top and made a steaming pot of coffee to have ready for her champion husband.

Sure enough, Mr Pepperpot soon came home—

without the cup. 'I forgot to take the wax,' he said, 'so I didn't think it was worth going in for the race. But I watched it, and you should have seen Paul Petersen today; I've never seen him run like that in all my born days. All the same, he looked very queer, as if he'd seen a ghost or something. When it was over he kept talking about his wife and his cap, and he wasn't satisfied till he'd telephoned his house and made sure his wife had been there all the time, watching the race on television.'

Then Mrs Pepperpot began to laugh. And ever since, when she's feeling sad or things are not going just right, all she has to do is to remember the day she went ski-racing in the wrong cap, and then she laughs and laughs and laughs.